HENRY PETROSKI

Invention by Design

HOW ENGINEERS GET
FROM THOUGHT TO THING

HARVARD UNIVERSITY PRESS

Cambridge, Massachusetts   London

Sixth printing, 2002

Library of Congress Cataloging-in-Publication Data
Petroski, Henry.
Invention by design : how engineers
get from thought to thing / p. cm.
Includes bibliographical references and index.
ISBN 0-674-46367-6 (cloth: alk. paper)
ISBN 0-674-46368-4 (pbk.: alk. paper)
1. Engineering design—Social aspects.
2. Engineering design—Political aspects.
3. Engineering design—Case studies.
I. Title.
TA174.P4735   1996
620′.0042—dc20
96-19227

To Stephen J. Petroski, E.I.T.

This book explores the nature of engineering and technology through case studies of familiar objects, from paper clips and aluminum cans to airplanes and modern high-rise buildings. These real-world artifacts (some of which I have written about before) are approached here from a perspective designed to illuminate different facets of the engineering enterprise—design, analysis, failure, economics, aesthetics, communications, politics, and quality control, to name but a few. The case studies also touch on a variety of engineering fields, including aeronautical, civil, computer, electrical, environmental, manufacturing, mechanical, and structural engineering.

I am grateful to the many designers, inventors, engineers, and entrepreneurs who have written to me about my earlier books and articles and who have shared their personal experiences. The case studies that I revisit in this book reflect new information gained from this correspondence—especially from inventors and engineers who participated in the development of the artifacts involved. Their elaboration of details has enabled me to tell more complete stories, to bring the material up to date, and to augment it with information about new and derivative products. I am also grateful to the many design theorists with whom I have corresponded and whom I have met in the course of traveling and lecturing. They have not only been gracious hosts but also have shared their references, insights, and approaches to the subject.

The rudiments of this book were first drafted in 1992 at Duke University in anticipation of teaching a new course employing case studies to introduce engineering students and others to the nature of invention, design, and development. I am grateful to the National Science Foundation for a grant that released me from teaching more traditional engineering courses at that time. Several classes of Duke students have now used various parts of the book and offered feedback that has guided successive revisions of the manuscript. I wish to thank especially Tonya Dale, Nayer El-Esnawy, and Ian Threlfall, who have served as teaching and research assistants for the course.

I am once again grateful to Albert Nelius, former head of circulation in Duke's William R. Perkins Library, and Eric Smith, former head of the Vesić Engineering Library. Dianne Himler has continued to secure essential library materials for me, and my graduate student, Aiman Kuzmar, while finishing his own dissertation in civil engineering, identified the materials that provided the background for the chapter on the fax machine.

Several editors and anonymous reviewers of an earlier version of the manuscript made suggestions that helped to shape its final form. I wish to express my sincere thanks to Harvard University Press's Michael G. Fisher, Ann Downer-Hazell, Susan Wallace Boehmer, and Jill Breitbarth for the enthusiasm, care, and help that have made turning the manuscript into this book a pleasurable experience.

As always, my daughter and son, Karen and Stephen, have been sources of inspiration. And my wife, Catherine Petroski, whose own writing I admire so much, continues to be my most cherished reader.

May 1996

CONTENTS

Invention by Design

1

The products of engineering are all around us. The computer on which these words were typed is an obvious example, as is the air conditioning system that keeps us (and our computers) comfortable even though it is hot and humid outside the building—which is also a product of engineering. When we do go outside physically, we often ride in an automobile on roads and highways with tunnels and bridges, and when we go outside metaphorically, we use telephones, videotapes, faxes, and computer networks. All of these are products of engineering design, manufacturing, and construction. Indeed, the world of our everyday experience is shaped by the practice of engineering and technology, and the world shapes those activities in turn. But what is engineering, what are its origins, and how do engineers practice it? And what is technology, what are its roots, and how does it relate to the rest of experience? This book explores such questions.

The process by which a high-tech computer gets from an inventor's brainstorm to our desk, or by which an innovative bridge gets transformed from some engineer's dream to a spectacular structure, is seldom straightforward. It can entail decades of painstakingly slow research and development, followed by weeks of frenzied activity. In addition, each engineering project is touched by the idiosyncracies of individual engineers, companies, communities, and marketplaces. And there are questions of economics, politics, aesthetics, and ethics. Furthermore, each engineering project is highly dependent upon the availability of raw materials of varying quality. And though engineering is the art of rearranging the materials and forces of nature, the immutable laws of nature are forever constraining the engineer as to how those rearrangements can or cannot be made.

This book attempts to make sense of many of these interrelated aspects of engineering. But rather than beginning with large structures or systems and the attendant mathematics and science that it takes to fully comprehend them, the book begins with a close look at some familiar objects. They may appear to be so simple as not to require any sophisticated engineering at all. But saying that these object do not require advanced engineering is not to say that these objects do not embody engineering principles. A self-taught and gifted inventor who comes up with some clever device and makes a million dollars on the idea uses thought processes not unlike those used by the graduate of an engineering school in designing and developing a new device to collect soil samples on Mars and bring them back to Earth. And as simple as familiar objects may seem at first glance, their conception, development, manufacturing, and marketing may pose great difficulties.

Engineering is a fundamental human process that has been practiced from the earliest days of civilization. Today, its methods have been professionalized and formalized, and its essential calculational nature has been greatly enhanced by the electronic computer. But that is not to say that the skills and discipline required to do good engineering are totally different from those exhibited by craftspersons throughout history. Modern engineering is a more highly mathematical and scientific endeavor, but its practice still requires a good deal of commonsense reasoning about materials, structures, energy, and the like. Mathematics and science help us to analyze existing ideas and their embodiment in "things," but these analytical tools do not in themselves give us those ideas. We have to think and scheme about nature and existing artifacts and figure out how they can be altered and improved to better achieve objectives considered beneficial to humankind.

The idea of design and development is what most distinguishes engineering from science, which concerns itself principally with understanding the world as it is. Engineers throughout history have wrestled with problems of water not being where it was needed, of minerals not being close at hand, of building materials having to be moved. Ancient engineers were often called upon to devise means of erecting great monuments, to design defenses against enemies, and to move people and goods across rough terrain and rougher water. By the Renaissance, engineers such as Leonardo da Vinci were coming up with more ideas than they could realize, and others such as Galileo were laying the foundations for the analytical methods that students now learn in engineering school. The Industrial Revolution brought a proliferation of new machines and manufacturing techniques, and this in turn provided an impetus for the growth of science and commerce on the international scale that we know today.

The history of engineering is rich with stories of success triumphing over failure; they are the basis not only for the technical but also the cultural background of engineering. Studying past cases of exemplary engineering, as superseded as their technology might be, can provide much insight into how we can approach some of the most challenging problems of today and come up with rewarding solutions. The stories of how engineers approached their problems and exercised their judgment have much to teach us about some of the fundamental, if not innate, characteristics of the human endeavor known as engineering.

CASE STUDIES

Case studies of artifacts or of particular products, projects, and processes enable us to understand engineering in the broad context in which it is actually practiced. While case studies must necessarily vary in their degree of detail and completeness, common features of engineering become evident as the stories reinforce one another and reveal aspects of the engineering method. Each case study presented here could itself be expanded into a book about what other alternatives might have been pursued or what improvements might still be made, for no engineering problem is ever completely solved to everyone's satisfaction. Engineering is the art of compromise, and there is always room for improvement in the real world. But engineering is also the art of the practical; engineers realize that they must at some point curtail design and begin to manufacture or build.

The paper clip, for example, is seemingly so simple and insignificant an object that we tend to use it while hardly noticing it. In Chapter 2, the paper clip is looked at under the microscope of technological criticism to see what it can teach us about design. In fact, the design and manufacture of a successful paper clip is an enormous challenge, as this case study reveals, and one that continues to this day. The most familiar paper clip, known as the Gem, is actually a far-from-perfect artifact, and since the Gem was introduced in the late nineteenth century inventors have patented many new designs. Looking more closely at the Gem and at some of the hundreds of patents for improvements on it provides insight into the nature of engineering design, and particularly into the question how an artifact that on close inspection is far from perfect can dominate the marketplace.

In Chapter 3, the case study of the pencil illustrates the means by which engineers analyze and quantify how things function in order to improve them. We all know that a pencil point will break if we press too hard upon it, but how hard is too hard? As users of pencils, we could try to play it safe and press very lightly so that our points would never break, but circumstances arise when pressing lightly will not be acceptable. Multiple-copy forms require us to print firmly, and to do so we must bear down on the pencil point. It is only after we break a few points off our pencil that we learn how to avoid breaking further ones. We can infer from our negative experiences how hard we can press on a pencil before it will break. We can also learn that a pencil point is less likely to break if we use a better pencil, if we hold the pencil more vertically, and if we use a blunter point. Alternatively, we could change our brand or type of pencil to one that has been improved so that it has a tougher point and does not break so easily. All such observations naturally have implications for larger engineering structures and systems as well.

The oftentimes long and arduous nature of development that follows the initial conception and even patenting of a promising device is exemplified in the story of the zipper, recounted in Chapter 4. The origins and history of this now ubiquitous slide fastener show the important role that financial backing and patience play in enabling a technical endeavor to proceed to commercial success. While the problem of developing a reliable zipper and manufacturing it in an economical fashion presented enormous technical challenges to the engineers involved, establishing a viable market for the product proved to be almost as elusive.

Another familiar artifact that puts a virtual engineering laboratory in our hands is the aluminum beverage can, the case study presented in Chapter 5. Understanding its origins and the constraints within which it has been developed helps us to understand, among other things, the key importance of failure criteria in engineering. The aluminum can was not developed as an end in itself but rather as a central component in the infrastructure of distributing beverages reliably in individual portions measured in the billions. When such large numbers are involved, the cost of each single unit plays a deciding role in the product's design.

Though economic considerations are of great importance, technological realities remain uppermost in determining how engineering shapes a product. For example, in designing a beverage can, not only must engineers fashion a can that will hold the beverage without contaminating it or allowing it to leak after rough shipping and handling, but they must also make the can easy to open and pour or drink from. Furthermore, while aluminum cans are a great convenience, they also represent an enormous potential waste of raw materials and energy, and getting rid of them raises significant issues of litter and waste disposal. Exploring the difficulties of satisfying numerous opposing goals and seeing how engineers and others have wrestled with such problems helps us to better understand the interplay between engineering, economics, and the environment generally. One aluminum can in isolation is one thing, but in the context of its billions of clones the aluminum can is quite another.

The story of the fax machine, told in Chapter 6, is the saga of a piece of sophisticated equipment that had its beginnings in the nineteenth century but could not develop into the user-friendly device it now is before a communications network over which data could be transmitted was put in place. The extraordinary rise in use of the fax machine that began in the 1980s had as much to do with government deregulation, standards development, and cultural imperatives as it did with pure technology. Engineering can never operate effectively in a vacuum.

Much engineering design and development that takes place in the modern context must proceed with a clear sense of worldwide developments, as the case history of commercial jet aircraft so effectively illustrates. As told in Chapter 7, the story of the Boeing 777—the twin-engine wide-body airliner that was introduced into passenger service in 1995— also provides an opportunity to examine the increasing role that digital computers now play not only in designing, testing, and manufacturing but also in operating some of the most complex machines and systems introduced in the late twentieth century. The enormous capacity and power demonstrated by the computer hardware and software used in such applications set the stage for engineering in the twenty-first century.

Not all engineers dream of designing large aircraft or other highly visible structures, for some engineers find the most elusive dreams in doing less ostensibly dramatic things more effectively, such as providing safe water supplies, disposing of waste, or cleaning up the environment. Because engineering is so inextricably involved with society and its goals, the practice of engineering is a very social endeavor, and this is perhaps nowhere so evident as in the case of water supply and disposal, presented in Chapter 8. In this field, even when the individual engineer is working alone at a desk, computer terminal, or drawingboard, every part of his or her work must potentially interface with everything else on a technical as well as a nontechnical level. In fact, no artifact or system that any engineer designs or analyzes can function independent of a larger social system, and the best designers and analysts are those who are constantly aware of the interrelationships of all things.

The story of just about any large bridge project can serve to illuminate the essential complexity of large projects, especially with regard to the political environment in which the engineering must proceed. In Chapter 9, the case history of the San Francisco–Oakland Bay Bridge shows the long gestation time typical of many large projects, as well as the numerous alternatives that can come to be proposed along the way. Issues of clearance for shipping, right of way for bridge approaches, rights of ferry operators, financial responsibility for construction, capacity of the bridge, aesthetics, and a host of different but interrelated problems experienced in the context of a specific bridge provide a paradigm for understanding great engineering projects generally.

In the late twentieth century, engineering has been greatly influenced by the introduction of systems of all kinds into virtually every aspect of every project, and the ancient art of building construction is no exception. The essential role that elevators and other systems have played in the evolution of the modern skyscraper is presented in Chapter 10. What might superficially be viewed as a structural engineering endeavor turns out to involve electrical, environmental, and mechanical engineering as well. Such interactions across traditional professional boundaries are ever-increasing and are sure to be a key component of twenty-first-century engineering.

By looking through engineering eyes at the things we encounter in everyday life, we come to see technological lessons in even the simplest of objects. Although we can admire products of elegant engineering and see them as examples to be emulated, it is also in being able to criticize poor engineering that we better comprehend how the made things of this world work and can be improved. A healthy sense of criticism is what gives the best inventors, designers, and engineers their edge, and it is also what drives the evolution of the made world and of technology. If we were content with everything around us, we would have no conception of improvement and the world would be a static place. While some might welcome that conservative approach, in which fewer risks were taken and fewer surprises would occur, it would also mean the curtailment of progress as embodied in the dreams of engineers, politicians, and people generally. Indeed, it has been held by some engineers that by not striving for more and more economical construction, the engineering profession would be irresponsibly appropriating limited resources to overdesign everything from beverage cans to bridges. While safe artifacts and systems of all kinds are clearly very important for the society that finances them, every dollar that goes unnecessarily into strength may be a dollar that is not available for other needs, such as conservation or maintenance. In the real world, such questions become inextricably intertwined with political and social issues, from which engineering problems are never immune.

2

A paper clip appears to be among the simplest of objects. In its most common form, it consists of a four-inch-long piece of wire shaped by three bends into a thing that is both pleasing to look at and easy to use. It comes fully assembled, and no batteries are required for its operation. No one expects instructions to come with a box of paper clips, and we tend not to think very much about how they are made and used. We take paper clips, as we do a lot of familiar artifacts, for granted, and seldom give them a second thought. They seem to be just too simple and ubiquitous to be very interesting or instructive. However, sometimes the simplest of things can hold as much mystery and provide as many lessons about the nature of engineering as the most complex.

When an object is simple and small enough to hold in our hand and turn about at will, we can inspect it to our heart's content to see for ourselves how it is made and how it works. If the artifact is inexpensive enough, we can each have an adequate supply to break open or test or experiment with in any way that might help us understand how the object is made and how it works. If the principles on which the object functions are conceptually simple and clearly visible, then we can explore questions of how we ourselves might engineer an improved version. Finally, the artifact can serve as a gripping metaphor for engineering itself.

Pick up a box of paper clips and examine it. The box is likely to have a minimum of information printed on it: The brand name (perhaps ACCO, which seems to be just another anonymous acronym, or Noesting, which seems to be an unpronounceable nonsense word); a name describing the kind of paper clips (perhaps Gem or Perfect Gem or Nifty or Peerless or Ideal—certainly something positive-sounding); the quantity in the package (usually a nice round number, like 100, but does anyone ever really count the number of paper clips in the box?); a catalog or stock number (so the supply can be replenished); possibly the address of the manufacturer (so the purchasing department knows where to reorder or locate a supplier, or complain about the product); most likely that ubiquitous UPC (universal product code) barcode that enables checkout counters to be automated; and, more likely a factor in selling the box than almost anything else printed on it, a picture of the kind of paper clips inside.

One thing that is *not* likely to be on a box of paper clips is instructions for use. We are all expected to know how to use these clever little objects, as readily as we know how to open the box and remove a clip, but we might be hard pressed to explain in words alone how to attach a single paper clip to a group of papers.

Let's open a box of paper clips and take one out, an action we are likely to perform without looking, letting our fingers select a clip from the top of the pile. We seldom, if ever, stop to admire or marvel at the paper clip. If our other hand holds the papers to be fastened, we may glance at the paper clip to see if it is oriented properly to slide onto the papers; if it is not, we will manipulate it around in our fingers without a thought. As we bring the clip to the papers, we will unconsciously notice the loops that must be slid one on each side of the papers. Experience will have taught us that a standard paper clip will not just slide automatically onto the papers, however; we must open the clip, most commonly by a rather subtle action of pressing the end of the longer loop against one side of the paper (which one of our fingers backs up and stiffens), while at the same time flexing the clip just enough so that it can be slid down on the papers with its smaller loop on the other side. This all takes place so quickly and automatically that the complex small motor skills required are usually overlooked, yet this action of applying the paper clip is central to its use—and to our appreciation of it as a piece of engineering.

THE SPRINGINESS OF MATERIALS

The paper clip works because its loops can be spread apart just enough to get it around some papers and, when released, can spring back to grab the papers and hold them. This springing action, more than its shape per se, is what makes the paper clip work. Springiness, and its limits, are also critical for paper clips to be made in the first place. To appreciate this, open a paper clip a bit wider than needed to get the loops around some papers. There will be a point at which the clip will be bent out of shape and will not return to the flat pattern that it had when fresh out of the box. When this happens, the clip's elastic limit is said to have been exceeded (or the wire is said to have been plastically deformed), and it is extremely difficult to restore the clip to the shape it had in the box. Needless to say, the clip is also no longer as effective in holding papers or in lying flat upon them.

Every material that engineers work with, whether it be timber, iron, concrete, or the steel wire of paper clips, has a characteristic springiness to it (not unlike the springiness of a rubber band), and the spring manifests itself in everything made of these materials. This behavior of materials was no doubt observed long before Aristotle's time, but it was a particular topic of discussion in that Greek philosopher's circle. In a collection of "mechanical problems" compiled in the fourth century B.C., the question was asked, "Why are pieces of timber weaker the longer they are, and why do they bend more easily when raised?" We have certainly all observed this behavior of long pieces of most everything: two by fours, spaghetti, pencil leads, plastic rulers, yardsticks, and so on. Anything slender can be bowed easily, and the longer the easier, yet if not broken or plastically deformed it will regain its straightness when put down. This is spring action, and it was not fully understood until about 2000 years after Aristotle's time.

Even the great Galileo did not fully recognize that all bodies have a certain springiness, and, as we shall see in Chapter 3, this led him to make some fundamental errors in his seminal work on strength of materials, published in 1638. It remained for Robert Hooke, a contemporary of Newton, to articulate the essential elements of elasticity. Hooke was an early advocate of the microscope and so was inclined to look closely at natural and artificial objects and to see things that other scientists overlooked. (Among the first observations Hooke reported in his *Micrographia*, published in 1665, related to the details of simple objects, such as the point of a needle or the edge of a razor.)

There was fierce competition among seventeenth-century scientists over priority of discovery of everything from calculus and natural laws to clever new devices, and so publishing a discovery in a cryptic manner established that one had made the discovery without having to reveal the details of it until the busy scientist or inventor had the time or inclination to do so in the way one would now, in the form of a scientific paper or a patent application. Although Hooke discovered the nature of spring force as early as 1660, he did not publish his observation about the elasticity or springiness of materials until 1678, and then in the form of an anagram.

The customary language of the time was Latin, and anagrams then did not have to spell out something apposite as they are expected to today (for example, THEY SEE is an good modern anagram of THE EYES). Thus, Hooke's anagram was presented with the letters in alphabetical order, as follows: *ceiiinosssttuv*. When he was ready to articulate his principle, he rearranged the letters to spell out, *Ut tensio sic vis*, which is commonly translated by the phrase, "As the extension so the force."

What Hooke had discovered was that, up to a limit, each object stretches in proportion to the force applied to it. Conversely, the more we stretch something elastic, the more resistance it offers to being further stretched. Thus if we pull a rubber band with twice as much force, it stretches twice as far. If we hold a very long piece of spaghetti by one end, it sags in a gentle but barely perceptible curve. Here the weight of the spaghetti itself is the force doing the pulling, and the stretching results from the bending that occurs. If the piece of spaghetti were too long, it could break, as it will if we cause it to vibrate, which adds the force of inertia to that of gravity and causes curving and bending beyond the Hookean, or elastic, limit. When the spaghetti or its broken parts are put back on the table, they are straight again, with the table providing support.

These springing phenomena are manifestations of Hooke's Law, and they (along with many other phenomena of materials and structures) affect the behavior of airplane wings and bridges and skyscrapers and paper clips and virtually everything mechanical and structural that engineers design. Heavy wire cables that support elevators in skyscrapers have a springiness that is exaggerated by the extreme length of the cable, and the bounce it produces can be unsettling to passengers if not properly taken into account in designing the elevator system.

A degree of elasticity can be very helpful in the operation of even the simplest of objects. If a straight pin, for example, did not have sufficient flexibility to allow it to bend a bit as it was threaded through a piece of fabric, the pin would tend to be more difficult to use and would not work as well. Furthermore, if it did not have enough spring and tended to stay crooked or plastically deformed whenever so slightly bent, it would not be so easily reusable. Although primarily intended to hold clothing together before buttons were commonplace, straight pins, like all technological artifacts, also came to be cleverly adapted for other uses. One important use was to attach papers together, long before anything like a modern paper clip was developed. To this day one can find pins used in this manner in third-world countries and in banking and brokerage businesses that cannot tolerate the risk of a paper clip slipping off documents

FIGURE 2.1 A mid-nineteenth century portrait of the British engineer Isambard King-dom Brunel, holding a pencil, with a wooden paper clip in the foreground

or the extra time it would take to remove staples. Before the metal paper clip, clothespins and other wooden clamplike devices were also used to fasten larger piles of papers together (as in Fig. 2.1), and in the mid-nine-teenth century the term "paper clip" more often than not meant a rather large metal clamp much like the kind that is found on clipboards today. But well into the twentieth century the straight pin remained a most common means of keeping a few sheets of paper together.

The kinds of wire used for centuries for making pins was also suitable for making paper clips, but the idea of a bent wire paper clip is more obvious in retrospect than in prospect. Bending wire into clever shapes is a very old concept, and even the Romans had such devices as safety pins. But before the advent of wire-working machinery in the mid- to late-nineteenth century, the process for making even a single pin or needle was rather long and tedious. In his famous book on the wealth of nations, the eighteenth-century Scottish economist Adam Smith used pin making as a prime example of division of labor and its economic benefits. Each step in the process was performed by a separate individual, and collectively they could make 4800 pins per day. Smith estimated that if an individual uninitiated to the techniques did all the steps on a single pin, the output might not even approach 20 pins per day. The famous French encyclopedia edited by Denis Diderot described pin and needle making in the late eighteenth century and illustrated parts of the process (Fig. 2.2). It was not hard to see the advantages of designing machinery to perform automatically all the tedious steps of pin making, but it was not until the 1830s that the American inventor John Howe succeeded in developing an effective pin-making machine (Fig. 2.3). Such a machine has been on

FIGURE 2.2 The very labor-intensive activity of needle making in the eighteenth century, suggesting also the division of labor involved in pin making at the time

FIGURE 2.3 A pin-making machine, patented by John Howe in the mid-nineteenth century

display in the National Museum of American History of the Smithsonian Institution, along with a videotape showing pins being made by it.

Paper clips could certainly be made by hand, just as pins had been for so long, but since pins served the purpose there was no pressing need to make such specialized objects as paper clips. With the development of the Industrial Revolution and the concomitant need to handle more and more volumes of paper as businesses expanded first nationally and then internationally, extremely specialized devices such as paper clips could be sold in such quantities as to make their manufacture worthwhile, if it could be done effectively.

Imagine how paper clips might be made by hand. They would most likely start much as a pin did, with a piece of wire pulled off a spool and straightened and cut to the appropriate length, say about four inches. This wire would have its spring, of course, but the fact that there was a limit to the spring would make the forming of a paper clip possible. As the wire was bent beyond the elastic limit, it would retain the bent shape. With the experience gained of trial and error, one could learn how to bend the wire, perhaps with the help of some pointed pliers, just far enough beyond the limit so that when the wire was released it would spring back to just the right shape. One could, after a while, develop a facility in doing this bending, and one could devise arrangements of pegs or jigs around which or in which to work the wire. In this way more paper clips could be made and, incidentally, made more quickly, and one might be able to make them inexpensively enough to sell for prices competitive with the pins they would displace in the office. Since pin making had become automated by the time wire paper clips were conceived, it essentially meant that paper clip making also had to be automated to produce a competitive product.

But paper clips could not have replaced straight pins on the basis of competitive price alone, and this brings us to one of the central technological ideas of invention and innovation and the roles that engineers play. A new artifact will displace an existing one only if there is a clear advantage that the new holds over the old. The most direct and successful means of establishing an advantage is to point out the shortcomings and failings of existing technology and to show how the new device serves to remove objections to the old. Nothing is perfect, and even the most traditional and established ways of doing things leave something to be desired. If a new artifact can be shown to overcome one or more incontrovertible disadvantages of an old, then there is likely to be some artifactual succession or evolutionary displacement. Generally speaking, however, the very fact that long-existing artifacts have become so familiar also means that people have adapted to any inconveniences or problems associated with their use. In fact, it is at first often only the inventor or engineer, effectively acting as technological critic, who even sees anything wrong with things as they are. But, once articulated, the problems that just one critic clearly points out will be immediately obvious to everyone.

If a new invention removes those problems, then it has a chance of succeeding.

The problems that late-nineteenth-century inventors found with the straight pin as paper fastener were several. It was difficult to thread through more than a few sheets of paper; it left holes in the paper; its point could prick one's finger; it could catch extraneous papers; it bulked up piles of paper. A flat paper clip that slid on and off a group of papers could be readily seen to do the job better by removing, or at least reducing, many if not all of these objections. Thus it was that early paper clips could displace pins in office use. But, as with many new products, early versions of the paper clip themselves soon came under criticism by other inventors. Early paper clips were generally not as easy to attach as subsequently conceived versions; early paper clips slipped off too easily; early paper clips got tangled together; etc., etc.

Whenever an inventor got an idea for a "new, improved" paper clip, its advantages were argued in contrast to the relative disadvantages of the old. A plethora of paper clip patents was issued around the turn of the century, but very few of the designs touted so successfully in the patent applications have survived. This is not surprising, since as each new artifact comes on the scene it becomes an object of criticism, especially by inventors who can imagine how this or that shortcoming (which, at first, only they see) can be removed, perhaps just by giving this or that leg of a paper clip a slightly different bend, turn, or twist. Not every inventor would choose to patent a new paper clip design, however, for various reasons. Some chose not to patent because the cost of the patent application seemed too high, others did not believe that the patent system was the best way to encourage invention, others felt they could maintain a better competitive advantage by keeping a trade secret than by revealing a new process in a patent application in exchange for the right to sue infringers.

THE GEM PAPER CLIP

For whatever reason, the most successful paper clip design, and the one that has become virtually synonymous with "paper clip," was never patented. Indeed, the concept of what has come to be known as the Gem clip clearly existed in the late nineteenth century because a patent was issued to William Middlebrook, of Waterbury, Connecticut, for a machine (Fig. 2.4) for *making* paper clips, and the patent drawings clearly show a fully formed Gem as the raison d'etre of the machine. Middlebrook's 1899

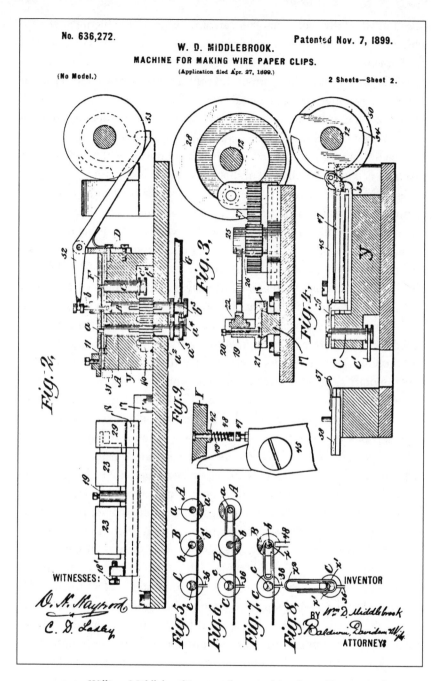

FIGURE 2.4 William Middlebrook's patent for a machine for making paper clips

patent incidentally shows that the standard history of the Gem paper clip, which credits its invention to a Norwegian named Johan Vaaler, is in fact not correct.

While Vaaler and other turn-of-the-century inventors were in fact patenting all manner of shapes and sizes of paper clips (see Fig. 2.5), Middlebrook was patenting the means for forming the Gem clip economically. Without machines like his, no paper clip could have challenged the machine-made pins very effectively. The complexity of Middlebrook's machine is clear from his patent drawings, and it is apparent that he was engaged in serious mechanical engineering as opposed to just doodling to find a new shape of paper clip. There could be many shapes of clip that can hold a pile of papers just about as well as, if not better than, a Gem, but the ability to manufacture the clips reliably and in large quantities is what would make or break a company. Why Middlebrook chose the Gem design to manufacture may be subject to speculation, but it clearly is the clip that the machine was designed to shape. The principles upon which the machine works, bending wire around pegs, are well suited to the Gem design and it to them. In short, Middlebrook's machine and the Gem were made for each other. But if not from Vaaler and his contemporaries, from where did the Gem come?

Believed to have been introduced in England in the late nineteenth century by a company known as Gem, Limited, the now-familiar paper clip that took its manufacturer's name soon became firmly established as *the* paper clip to which all others were compared. Patents continued to be issued for improvements in paper clips, but none displaced the Gem. Today, when an icon for the paper clip is needed to label a computer's desktop organizer or to warn against jamming copier machines, it is the Gem that seems invariably to be employed. Newer clips, such as the plastic-coated wire variety, are shaped like Gems, although their proportions never seem to be quite right. One of the latest genuine improvements in (Gem) paper clips is the introduction of a turned-up lip on the end of the inner loop. This enables the unopened clip to be truly slid on the papers, and there is no need for the user to spread the loops of the clip apart manually. But, as with just about all improvements, there is a trade off. This new clip is not flat, and so it adds further bulk to a pile of papers.

In fact, this new improvement is not even all that new, for it was clearly present in at least one version of paper clip patented in 1903 by George McGill of Riverdale-on-Hudson, New York. Several versions of his clip, as illustrated in the patent drawings (Fig. 2.6), show clearly that McGill also sought to improve the Gem by turning up or putting an eye or bulb on the ends of the wire, which tended to catch and rip the paper upon removal, as well as to introduce a turned-up end on the inner loop to make application easier.

IMPROVEMENTS IN PAPER CLIPS

Inventors are always looking for things to improve, and for about a century the Gem has been the main target of criticism in patents for new and improved paper clips. For example, one clear challenge to the Gem was patented in 1934 and has come to be known as the Gothic clip (Fig. 2.7), because its loops are pointed more to resemble Gothic arches than the rounded Romanesque ones of the Gem. Henry Lankenau's patent application for the "perfect Gem" also listed ease of applying to papers as one of the invention's advantages. More importantly, the Gothic clip has longer legs that extend almost to its squared end, thus reducing the possibility that their sharp ends would catch and tear paper. Since the danger of tearing papers or the pages of books is minimized with this clip, it can typically be made of heavier wire to give it better gripping power. While it is also more expensive, the Gothic clip is favored by some users, such as librarians, because of its distinct advantages.

There are other ways to improve the paper clip, and among the most often tried is economizing on raw materials, a common object of engineering design and manufacturing. After the capital investment that goes into the machinery to make paper clips, the wire that is used is the single most controllable factor in determining cost and hence price. Because invention, design, engineering, and manufacturing are always bound by the laws of nature, only so much savings can be effected by reducing the quality of the wire used. There must be just the right spring to the paper-clip wire, and to try to make clips with too stiff or too soft a wire is tantamount to trying to break Hooke's Law. Material economies can be realized in another way, however, and that is in the amount of wire used in each paper clip. Starting with a piece of wire just ten percent shorter that what the competition uses to fashion its Gems can translate into an advantage in the office products catalog, especially if saving pennies on every box of paper clips is more important than how the clips look to a supply manager who orders them by the millions.

The classic Gem paper clip has certain proportions, as shown in Fig. 2.8, which are principally manifested in the distance between the inner

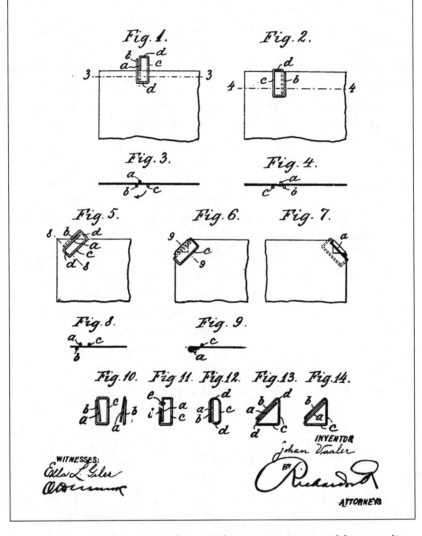

FIGURE 2.5 Two of the many early twentieth-century patents granted for paper clips

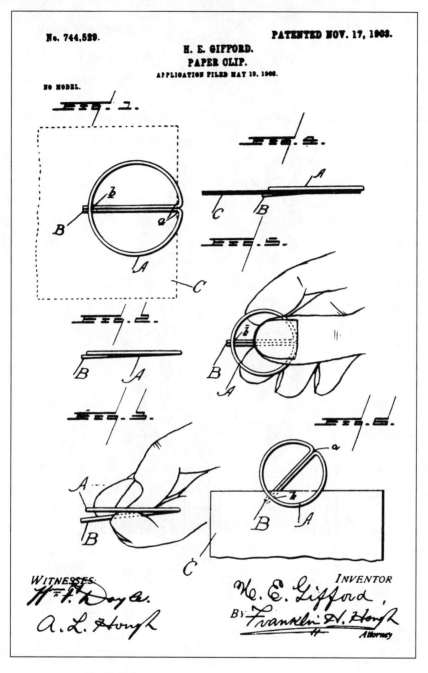

FIGURE 2.5 *(Continued)*

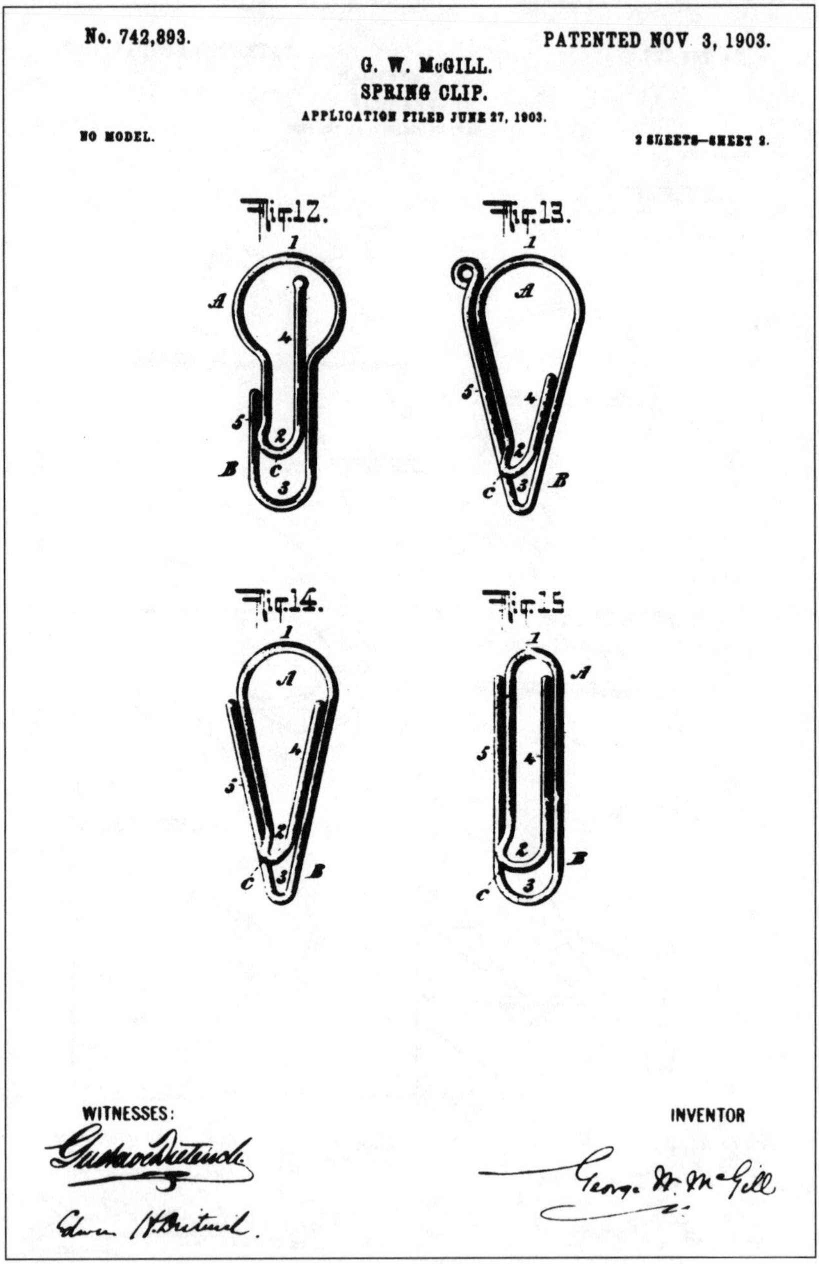

G. W. McGILL.
SPRING CLIP.
APPLICATION FILED JUNE 27, 1903.

NO MODEL.

2 SHEETS—SHEET 2.

Fig.12.

Fig.13.

Fig.14.

Fig.15

WITNESSES:

INVENTOR

George W. McGill

FIGURE 2.6 An early twentieth-century patent for a wide-lipped paper clip

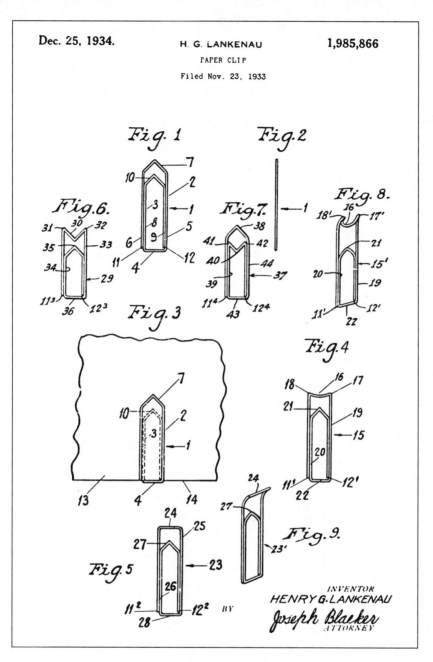

Dec. 25, 1934.   H. G. LANKENAU   1,985,866

PAPER CLIP

Filed Nov. 23, 1933

FIGURE 2.7 Patent for a Gothic-style paper clip, issued in 1934

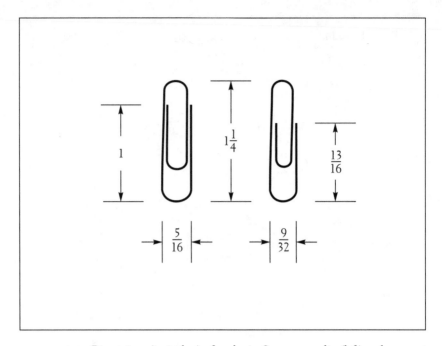

FIGURE 2.8 Dimensions (in inches) of a classic Gem paper clip (left) and a recent imitator

and outer loops and the length of the legs in which the wire terminates. The longer the inner loop, and thus the closer it is to the outer loop, the harder it is to attach a standard Gem, but the longer the loops and the legs, it has been generally believed by inventors, the better the gripping action of the clip. Like many a problem in engineering design, there is a tradeoff no matter what proportions are chosen, and the "best" solution is the one that balances ease of application with gripping power. The compromise might be further complicated if the designer wishes to choose balanced proportions that make the clip also aesthetically pleasing to the eye.

The proportions of the Gem illustrated in Middlebrook's 1899 patent for his machine seem to have struck the happy medium that gave an all-around successful paper clip. Like all artifacts, the Gem had its shortcomings, but for almost a century its proportions remained unchanged. When old companies wanted to make a less expensive Gem or newer manufacturers wanted to break into the market, one of their most obvious strategies was to use less wire per clip, which necessitated making judgments about how to reshape the classic lines of the Gem. A proportionately smaller clip would be one solution, but the overall length of the clip would then change, and that would make the clip seem not quite comparable to the standard Gem. Alternatively, the loops could be moved further apart or the end legs could be shortened, or thinner wire could be used, or all of the above could be incorporated in a very inexpensive paper clip. Many of the newer manufacturers of "Gems" being sold today have taken this last course, making the clips look, feel, and behave differently. Each such alteration also leads to functional problems, of course, such as reduced gripping power, more easily deformed clips, and more easily torn papers.

CHANGE AND COMPETITION

The box of "No. 1 Gem Clips" that I took most recently from our office supply cabinet displays a drawing of several Gems on the cover. These are classically proportioned, with close (but not too close) loops and long straight ends. The clips inside the box are different, however, as can easily be seen by placing one of them over the cover drawing. The actual clips, formed of less wire, have more distant loops and shorter legs. This gives a design that appears to be distinctly less well proportioned. In examining the box more carefully, I discovered that the clips are imported from an emerging industrialized country, and it is very possible that they were made on reclaimed old machinery that is not too dissimilar from Middle-brook's but that has been adjusted to cut off a shorter piece of wire before forming it into an almost but not quite classically proportioned paper clip.

Besides the poor proportions of the clips in this box, there is an amazing lack of uniformity from clip to clip, contrary to all the conventional wisdom about mass produced items. Although the side of the box reads "Finest Quality, Bright Finish, and Smooth Style," the contents proved to be a mixture of bright and dull finish clips, smooth and ridged designs, and, most strikingly, a collection of clips that looks almost hand-formed, with some so distinctly malformed that they could hardly be called paper clips. (In one box, I even found a straight pin, recalling the origins of the paper clip and the similarity of the roots of the manufacturing process.) Few of the clips lay flat, many are bent out of their plane and look almost pre-used. Clearly, there was little attempt to control the quality of what goes into the box. Since so few of the clips in the box are flat or have close-set legs, they tend to get tangled together and thus it is difficult to pick out one without several others dangling from it. This problem was described in some of the earliest patents for paper clips, and it was considered such a distinct disadvantage almost a century ago that many a new paper clip design argued for its superiority by claiming that the clips did not become interlocked in the box. Once the faults are pointed out, it is easy to see that the "new, improved" paper clips in the supply cabinet are in fact inferior to the old, established brands. So how did the new brand of paper clips end up in the supply cabinet? Why do manufacturers risk producing products inferior to what already exists?

A book on "successful product design" provides some insight. Written by a former director of design in a manufacturing company and by a lecturer on technical management, the book outlines a concept of "total design," whereby technical issues of design and production are balanced by economic and marketing considerations. As an illustration of the system that the authors have devised, they present a detailed case study of the hypothetical British company, Omega Wire Forming, a large firm that had long prospered by supplying the automotive industry with springs and wire clips. With that industry depressed, however, Omega wished to broaden its product line so that it could move into new markets. The Board of Directors ordered the firm's design manager to form a group consisting of members of the market research, production, sales, and design departments and to look into what new product could be made utilizing the company's experience in wire forming. There were naturally guidelines on how much capital might be invested in new machinery and how much return on investment was expected, but otherwise the research group was unconstrained.

Within two weeks, the group identified wire paper clips as a viable new product. The decision was justified in part by data showing that in recent years the price of paper clips had risen almost three times as fast as the price of wire, that demand for the product was up, and that there appeared to be little customer loyalty to any particular brand. According to the group's analysis, if Omega could manufacture paper clips at three for a penny, it might expect to capture 10 percent of the domestic market, which was estimated to be 500 million paper clips per year. Furthermore, if the price could be lowered to four for a penny, Omega might expect to sell 80 million paper clips domestically and export another 20 million each year. Among the main anticipated costs was that of distribution.

Omega Wire Forming decided to get into the business of making paper clips, and the company's designers and engineers were instructed to focus their attention on improving the Gem paper clip and its method of manufacture. Since the standard-size Gem accounts for 90 percent of the market, its overall dimensions were believed to be fixed. However, if the legs could be cut off shorter, up to 10 percent less total wire would be needed for each paper clip, thus reducing the cost of raw materials, which represents over half the manufacturing cost. A lighter paper clip would result, which would also reflect savings in shipping costs. A test engineer's determination that a paper clip with shorter legs resulted in only a 2 percent decrease in springiness or gripping strength clinched the decision.

Finally, at virtually no additional manufacturing cost, the inner loop of the paper clip could be given a slight bend out of its plane, a competitive feature that had recently begun to appear on most competing British paper clips, thus reviving a feature that the inventor McGill had introduced almost a century earlier (see Fig. 2.6). Within a year of studying the problem, Omega Wire Forming was manufacturing modified Gem paper clips and was expecting to recover its investment costs in eighteen months. Thus are the economics of saving fractions of a penny on items that have potential sales in the hundreds of millions, if not billions worldwide.

Whether the Far East manufacturers of the new paper clips found lately in our stationery catalogs and supply cabinets followed so structured a total design process as that of Omega Wire Forming, the imported clips do embody some very similar cost reduction features to give them an edge over more traditional domestic designs. It behooves those older manufacturers who do not wish to lose very much more of their accustomed market share to the newcomers on the shelf to look at the traditional designs with a critical eye to improving their usability and to educating paper clip users about the aesthetic, strength, and quality advantages that might be had for pennies more a box. The engineer, often acting as an inventor or designer, plays a central role in such considerations, whether the product be paper clips or computer chips. Indeed, the engineer is often the most severe critic of existing technology, and that is why things change over time. Curiously, however, where engineers find fault and see a need for improvement, others can sometimes find perfection.

FORM AND FUNCTION

The classic Gem paper clip is often held up by product designers as the epitome of modern manufactured articles. One design critic, praising elegant design solutions, has written:

> If all that survives of our fatally flawed civilization is the humble paper clip, archaeologists from some galaxy far, far away may give us more credit

A favorite pastime of some office workers is to doodle in wire by reshaping paper clips into all sorts of fanciful, and sometimes grotesque, new forms.

Try your hand at deconstructing a Gem and designing a new paper clip. How is your design an improvement on the Gem? Does it have any less desirable qualities, such as reduced gripping power? Inventors often claim their improved designs for paper clips have superior gripping power to that of the prior art. How could you determine in an objective way which of two paper clips has the greater gripping force under comparable conditions?

than we deserve. In our vast catalog of material innovation, no more perfectly conceived object exists. To watch a Middle Eastern bank teller bind together bank notes with a straight pin (a practice dating back to early Roman times) is to comprehend the paper clip's amazing grace.

And an architecture critic, marveling at "commonplace things of great design," clearly had the Gem in mind when he wrote:

> Could there possibly be anything better than a paper clip to do the job that a paper clip does? The common paper clip is light, inexpensive, strong, easy to use, and quite good-looking. There is a neatness of line to it that could not violate the ethos of any purist. One could not really improve on the paper clip, and the innumerable attempts to try—such as awkward, larger plastic clips in various colors, or paper clips with square instead of rounded ends—only underscore the quality of the real things.

These excerpts are typical of the kind of praise heaped upon the Gem paper clip by those who look more at form than function. In their original context, such quotes are accompanied by illustrations of one or more Gems, but almost never clipping papers together. While the Gem may be a truly attractive and elegant object to look at when photographed in its own right, when applied as intended to a group of papers it takes on a less graceful appearance. The "bravura loop-within-a-loop design" is broken by the papers it is intended to hold, and the Gem can look curiously incomplete and misaligned, with its straight end not on target to meet the loop disappearing behind the sheets. When applied to more than a few sheets, the Gem can appear especially ungainly, twisted, and deformed.

Engineers and inventors are not so easily pleased with the object in the abstract. While they are not averse to designing things that look attractive, that is not the only criterion for elegance and beauty, which can only be skin deep. How a thing functions is often where engineers begin their quest, and no object that fails to function properly can be considered truly beautiful or perfected. To a host of inventors, the perfect paper clip remains an unfulfilled quest, as the steady stream of patents for new, improved paper clips attests. Even as the one-hundredth anniversary of the Gem approached, patent applications continued to be filed for still more improvements on the Gem and other paper clip designs.

Among the imperfect things about the Gem that many a recent inventor has discovered or rediscovered when reflecting upon how the "perfected" paper clip is used to clip papers together are the following:

1. *It goes on only one way.* Half the time, the user has to turn the clip around before applying it.
2. *It does not just slip on.* The user has first to spread the loops apart.
3. *It does not always stay on.* The clip gets snagged on papers or other objects and gets pulled off.
4. *It tears the papers.* The sharp ends of the clip dig into the papers when it is removed.
5. *It does not hold many papers well.* The clip either twists badly out of shape or flies off the pile.
6. *It bulks up stacks of paper.* A lot of file space can be taken up by paper clips.

When a new design removes one of these annoyances, it more likely than not fails to address some others or adds a new one of its own. This is what makes engineering and inventing so challenging. All design involves conflicting objectives and hence compromise, and the best designs will always be those that come up with the best compromise. Finding a way to bend a piece of wire into a form that satisfies each and every objective of a paper clip is no easy task, but that does not mean that people do not try.

A MODEL PAPER CLIP PATENT

Over the past century there have been several hundred patents issued specifically for bent wire paper clips, and each of them is a testament to its inventor's ability to find fault with existing paper clip designs, which more often than not are represented by the Gem. That the quest remains as challenging as ever can be demonstrated by looking at some paper clip patents from very recent years.

U.S. Patent No. 4,949,435 was issued on August 21, 1990, to Gary K. Michelson, a surgeon and inventor of medical devices from Venice, California. The cover page is in the format that has been followed since the early 1970s, when the U.S. patent and trademark laws were revised (see Fig. 2.9). This new format shows various things at a glance, including the patent number, name of the patent, name of the inventor, application filing date, which is not untypically a year or more before the patent is granted, and the date of the formal issuance of the patent, which by tradition will be at noon on a Tuesday. (In most countries, patent rights go to the person who files first. In the United States, the law holds that patent rights belong to the person who first came up with the idea, and

United States Patent [19]

Michelson

[11] **Patent Number:** **4,949,435**

[45] **Date of Patent:** **Aug. 21, 1990**

[54] **PAPER CLIP**

[76] Inventor: **Gary K. Michelson**, 438 Sherman Canal, Venice, Calif. 90291

[21] Appl. No.: **257,849**

[22] Filed: **Oct. 14, 1988**

[51] Int. Cl.⁵ B42F 1/02
[52] U.S. Cl. 24/67.9; 24/546
[58] Field of Search .... 24/67.9, 67 R, 67.3, 24/67 CF, 545, 546, 547, 548, 549, DIG. 8, DIG. 9, DIG. 10; D19/65

[56]   **References Cited**

U.S. PATENT DOCUMENTS

184,626	11/1876	Jewett	24/546
395,473	1/1889	Bartley	24/67.9
715,992	12/1902	Cox	24/548
743,017	11/1903	McGill	24/545
795,048	7/1905	Maguire	24/67.9
1,334,233	3/1920	Dinwiddie	24/547
1,336,626	4/1920	Hall	24/547
1,783,099	11/1930	Ries	24/546
2,642,638	6/1953	Larrabee	24/67.9
2,822,593	2/1958	Sponsel	24/67.9
4,286,358	9/1981	Levin	24/67 R

4,665,594	5/1987	Wagner	24/546

FOREIGN PATENT DOCUMENTS

317844	9/1902	France	24/67.9
1439151	4/1966	France	24/370
709353	5/1954	United Kingdom	24/67.9

OTHER PUBLICATIONS

Horders Inc. Cat. #56, 1952, One Sheet "Paper Clips and Fasteners".

Primary Examiner—Victor N. Sakran
Attorney, Agent, or Firm—Lewis Anten

[57]   **ABSTRACT**

An improved paper clip is disclosed consisting of a single piece of wire bent so as to have a straight top spine portion, two straight side leg portions substantially perpendicular to each end of the top spine portion and bent portions extending from, each side leg portion diagonally at approximately 45 degrees in the direction of the top spine portion. Each bent end portion extends from more than ⅛ the diagonal distance from the end of the side leg portion to the top spine portion.

1 Claim, 2 Drawing Sheets

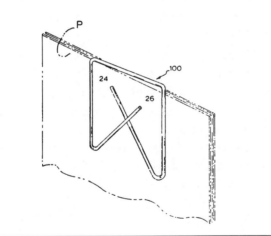

FIGURE 2.9 Patent issued to Gary Michelson for a new paper clip

this is why so many industrial and individual inventors scrupulously keep carefully dated notebooks of their ideas, experiments, and explorations.) Also on the cover page are lists of Patent Office categories known as classes and fields involving the so-called prior art, or existing relevant technology, against which the novelty of the improvement is judged by the patent examiner. Related patents, both U.S. and foreign, are also listed on the cover page, as are other (nonpatent) documents that might be relevant to the case. The cover sheet also shows the name of the patent examiner (a job that Albert Einstein held in Switzerland while he was formulating his theory of relativity) and the name of the inventor's patent attorney, agent, or firm. Finally, there is an abstract that briefly describes the invention, the number of claims (which appear on the last page of the patent and which spell out for what exactly the inventor has been granted a patent), and the number of drawing sheets (which can show the prior art as well as various views of the subject of the patent).

Drawings are a very important component of the patent, and an application can be delayed until drawings acceptable to the Patent Office are produced. The sheets of drawings follow the cover page, and in the case of Michelson's paper clip patent there are two sheets, or pages, of drawings, showing the paper clip from various points of view, with its various parts numbered for reference in the text of the patent. The drawing appearing on the cover page is one of those that appears on the inside sheets also. While patent drawings are very important in conveying the idea of the patent, these drawings do not always follow the best practices of engineering drafting, and so the device might not be easily understood from the patent drawings alone.

The textual part of the patent document begins with background to the invention, and the opening paragraphs of Michelson's patent are especially explicit in criticizing the prior art. Even though some of the language might be a bit excessive, as it is wont to be in patent applications, there should be little doubt that he is talking about the Gem:

> Paper clips are generally used as expendable items to either temporarily attach papers or as indexing markers.
> Use of the common paper clip presents a number of problems. In use, the two loops of the common paper clip must be digitally manipulated and manually spread so that papers can then be inserted between the loops. The common paper clip does not work well for any significant thickness of paper, being difficult to apply, and once applied is physically deformed, in the process, such that it cannot be reused without being deliberately

and correctionally bent by the user. Further, when used with any significant thickness of paper, the loops of the common paper clip torque on their long axises [sic] such that the clip will not lie flat on the papers and will simultaneously protrude from the plane of the papers on either side. This also results in the sharp ends of the legs digging into the papers causing damage. More damage will usually occur during removal of the clip as the clip is dragged even deeper into the paper.

Also, because the loops of the paper clip are being spread apart, there is a tendency for the common paper clip to try to close by forcefully ejecting off of a thick group of papers, thereby presenting a very real threat to the user, who may be struck in the eye.

Michelson goes on in his patent with further criticism of the Gem when used as an indexing marker, such as it might be in marking the page of a book. The patent document then describes the new and improved paper clip, and Michelson is especially explicit in showing point by point how his paper clip is, "despite its relative apparent simplicity, a technological advancement and is superior to the prior art" with regard to his categories of capacity, reusability, ease of application, ability to lie flat, ease of removal, indexing features, safety, economics, and mailability.

The form of patents involves a lot of redundancy, and Michelson's patent is no exception. It goes on to state the object of the invention, which is essentially a rehashing of the description of the clip's advantages. Separate brief and detailed descriptions of the drawings follow, referring to the little numbers associated with the various parts of the device illustrated. Finally, the patent closes, as does every patent, with a legally-crafted single long sentence, filling an entire paragraph, and prefaced by the phrase, "What is claimed is." It is this final section that is the ultimate focus of the patent examination, for it is what the applicant is allowed to claim or not for his or her invention that provides full protection or none. If none of the applicant's claims is allowed, then no patent can be issued.

The single claim of Michelson's patent is about 200 words long, and it begins:

What is claimed is:
1. A paper clip comprising a single piece of bent wire having a straight top spine portion, two side leg portions bent substantially perpendicular to the top spine portion and two bent end portions bent substantially at a 45 degree angle to the side leg portions . . .

Another new paper clip was invented by Charles T. Link, of Carmichael, California, and on November 12, 1991, he received U.S. Patent No. 5,063,640 for his "endless filament paper clip," as shown in Fig. 2.10. Link's principal focus was the lack of symmetry of common paper clips like the Gem. When we pick a Gem out of a box, we have to make sure it is turned the right way before attaching it to a group of papers. This failing is addressed by the endless filament clip, which can be applied from either of its ends. Rather than go into his patent in detail, however, it will be instructive to look at Link's inventive process. When he was asked how he came to invent his clip, Link wrote in a letter about three aspects of invention:

1. Why did I want to invent the clip? . . . I had been convinced for many years, and others whom I talked to concurred, that people would welcome an improvement in paper clips that would enable one to engage papers in the same manner at both ends of the clip, and that would also eliminate scratching and tearing of the papers. The more I used paper clips the more convinced I became that these improvements should be made and that the effort might be financially rewarding.

2. Why did I believe I could invent it? . . . My initial effort to sketch a likely concept frankly cooled my enthusiasm somewhat. . . . After further days of sketching and thinking, however, a rather vague concept began to materialize and gradually bring about a reasonable confidence in my mind that I could probably create a satisfactory configuration for a new clip having the desired improvements.

3. How did I invent it? . . . The third stage took place over a much longer period of time because it included not only sketches and drawings but also considerable experimenting with variously modified configurations of clips and with various means for forming them. This period also included a great amount of time and effort spent in collaborating with my patent attorney . . . to produce the best configuration, claims and drawings we could.

Gaining a patent does not end an inventor's quest, however. Link has also written of the difficulties in locating a wire fabricating concern that could produce the clip. A special wire-forming machine may have had to be invented. In the meantime, Link made sample clips "by hand with the aid of a special wire bending fixture." The clips are very attractive and a pleasure to use, but whether they would ever be available in stationery

United States Patent [19]

Link

[11] **Patent Number:** **5,063,640**

[45] **Date of Patent:** **Nov. 12, 1991**

[54] **ENDLESS FILAMENT PAPER CLIP**

[76] Inventor: **Charles T. Link, 6227 Pattypeart Way, Carmichael, Calif. 95608**

[21] Appl. No.: **604,085**

[22] Filed: **Oct. 26, 1990**

[51] Int. Cl.⁵ B42F 1/02; A44B 21/00
[52] U.S. Cl. 24/67.9; 24/67.3; 24/552
[58] Field of Search .... 24/67.9, 67.3, 67 R, 24/546, 556, 550, 552

[56]   **References Cited**

U.S. PATENT DOCUMENTS

1,053,008	2/1913	Carbis
1,251,884	1/1918	Hann 24/552
1,271,043	7/1918	Lee 24/552
1,767,973	6/1930	Gedney 24/552
2,061,474	11/1936	Metzs 24/552
2,074,613	3/1937	Laursen 24/552
2,152,075	3/1939	Melahn 24/552
2,239,584	4/1941	Zane 24/552
4,170,052	10/1979	Okerblom	
4,480,356	11/1984	Martin 24/67.3

Primary Examiner—Victor N. Sakran
Attorney, Agent, or Firm—James M. Ritchey

[57]   **ABSTRACT**

A paper clip for securing to at least one sheet of thin material comprises a continuous or endless filament shaped to provide two oppositely facing pincers. Each oppositely facing pincer has a resilient hinge region connecting and urging together an opposing pair of pincer finger members with each pincer finger member comprising a tip segment, an inner gripping segment, and an outer edge segment. The sheet of thin material is secured between either of the oppositely facing pincers by the inner gripping segments of the respective pair of opposing pincer finger members.

20 Claims, 4 Drawing Sheets

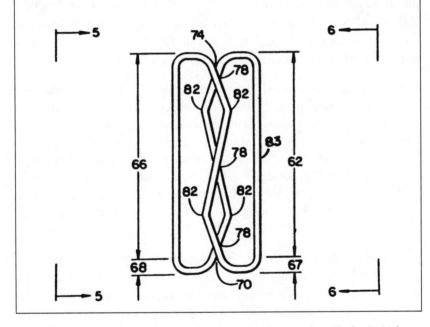

FIGURE 2.10 Patent for an endless filament paper clip, issued to Charles T. Link

stores would depend upon how successfully made and marketed they could be in the minds of manufacturers and investors. Link has written reflectively of his design experience as follows:

> In retrospect, the visual aspect of the overall design problem seemed to be at least as difficult and frustrating as the original functional problem, partly because of the subtle nature of the former and partly because of the need to achieve visual compatibility with the billions of Gem clips in use. In addition, however, the visual aspect seemed to be sensitive to the passage of time which frequently resulted in an apparently excellent visual design change losing its appeal some time after being introduced. Then again, some visual design changes seemed to gain in appeal over time. Fortunately, the visual and functional aspects of the entire clip design finally reached compatibility and stability over time.

Another contemporary inventor, Billie E. Strong, of Midpines, California, also addressed the Gem's failure to be applied from either direction. U.S. Patent No. 5,170,535 was issued to Strong on December 15, 1992, for "time saving paper clips," as shown in Fig. 2.11. In the background to the invention, Strong criticizes the Gem as follows:

> In reviewing this standard form of paper clip, it will be noted that only one end of the clip can be utilized to clip pieces of paper together. As such, when a paper clip is removed by a worker for the purpose of clipping sheets of paper together, the worker must first determine the correct end of the clip and align it accordingly. This greatly increases the task of clipping sheets of paper together, both in time and effort, and accordingly, there exists a continuing need for new and improved paper clips which could be more easily aligned and utilized in performing this simple office function.

Like virtually every inventor, Strong wants to respond to this "need," which admittedly may not be as readily perceived by the typical office worker, by providing "a new and improved paper clip which has all the advantages of the prior art paper clips and none of the disadvantages." The new clip has the added advantage that it can hold a label and so also serve as a filing aid.

Still another inventor, Suzy Chung Hirzel, found fault with large plastic paper clips because they could not hold thick piles of paper, and she was issued U.S. Patent No. 5,010,629 on April 30, 1991, for her "paper clip with vertical panel," as shown in Fig. 2.12. According to this inventor, who markets a variety of her own products under the name Suzy Chung

United States Patent [19]

Strong

||||||||||||||||||||||||||||||||
US005170535A

[11]	Patent Number:	**5,170,535**
[45]	**Date of Patent:**	**Dec. 15, 1992**

[54] **TIME SAVING PAPER CLIPS**

[76] Inventor: Billie E. Strong, 5275 Davis Rd., Midpines, Calif. 95345

[21] Appl. No.: 830,927

[22] Filed: **Feb. 4, 1992**

[51] Int. Cl.⁵ B42F 1/00; G09F 3/00
[52] U.S. Cl. .... 24/67.9; 24/67 R; 24/547
[58] Field of Search .... 24/67.9, 67 R, 67 AR, 24/545, 546, 547, 549, 90 HA

[56] **References Cited**

U.S. PATENT DOCUMENTS

1,418,306	6/1922	Holt	24/547
1,516,294	11/1924	Hubeny et al.	24/547
1,783,484	12/1930	Ross	24/67.9
2,055,152	9/1936	Larson	24/547
2,781,566	2/1957	Hammer	24/67.9
2,815,595	12/1957	Davis	24/67 AR
2,823,479	2/1958	Zdanowski	24/90 HA
3,225,469	12/1965	Chase	24/67.9
3,408,700	11/1968	Chase	24/67.9
3,913,181	10/1975	Walker	24/67.9
4,237,587	12/1980	Hsiao et al.	24/67.9
4,286,358	9/1981	Lenin	24/67.9
4,300,268	11/1981	Wilson	24/67.9
5,022,124	6/1991	Yiin	24/67.9

FOREIGN PATENT DOCUMENTS

0444860	5/1927	Fed. Rep. of Germany	24/547
1128624	1/1957	France	24/67.9
0281894	12/1927	United Kingdom	24/67.9

Primary Examiner—Victor N. Sakran
Attorney, Agent, or Firm—Leon Gilden

[57] **ABSTRACT**

An improved paper clip is provided with an extra loop of wire so that the clip can be used from either end to attach papers together. Modifications include providing the clip with serrated gripping edges, bend points and adhesively attachable labels.

2 Claims, 4 Drawing Sheets

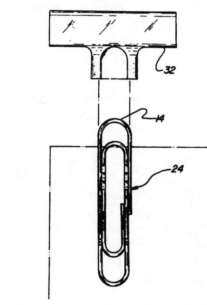

FIGURE 2.11 Patent for a time-saving paper clip, issued to Billie E. Strong

United States Patent [19]

Hirzel

[11] **Patent Number:** 5,010,629

[45] **Date of Patent:** Apr. 30, 1991

[54] **PAPER CLIP WITH VERTICAL PANEL**

[76] Inventor: Suzy C. Hirzel, 933 Shellwood Way, Sacramento, Calif. 95831

[21] Appl. No.: 457,948

[22] Filed: Dec. 27, 1989

[51] Int. Cl.⁵ B42F 1/02
[52] U.S. Cl. .... 24/67.9; 24/547; 24/67 R
[58] Field of Search .... 24/67.9, 67 R, 67.3, 24/67.11, 3 J, 545, 546, 547, 549

[56] **References Cited**

U.S. PATENT DOCUMENTS

1,809,689	6/1931	Graves	24/549
1,972,434	9/1934	Yerk	24/3 J
2,184,569	12/1939	Stewart	24/67.9
2,768,416	10/1956	McMullen	24/547
2,910,749	11/1959	Parker	24/67.9
2,938,252	5/1960	Scheemaeker	24/547
3,168,954	2/1965	von Herrmann	24/67 R
3,914,824	10/1975	Purdy	24/67.9
4,332,060	6/1982	Sato	24/67.9

FOREIGN PATENT DOCUMENTS

568514	6/1958	Belgium	24/67.9
978127	4/1951	France	24/67.9

Primary Examiner—Victor N. Sakran

[57] **ABSTRACT**

An improved clip with a neck that enables user to easily secure the thick gathered papers or the like without wrinkling or cramping. The clip having a vertical neck at the back of the clip, between the upper grip and lower panel, and lower panel is larger than the upper grip. Lower grip is flat and straight but upper grip is slanted. Front tip of the upper grip is almost the same level as the front tip of the lower panel and rear part of the upper grip is the same height as the top of the neck. The clip also has a rear solid portion having a back edge adapted to be pushed down by fingers. The clip will allow one to insert approximately ¼" to more than ½" papers or like material.

The shape of the clip is generally round at the front tip and generally straight at the rear end which is a handle portion. The upper grip may have an additional smaller grip and it can/may hold an additional layer of the paper or like.

The clip is made of one piece by plastic by injection molding and stamped metal sheet, or single wire.

5 Claims, 2 Drawing Sheets

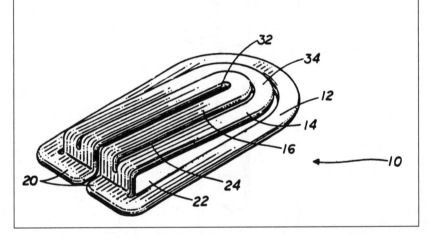

FIGURE 2.12 Patent for a large-capacity plastic paper clip, issued to Suzy Chung Hirzel

or simply Suzy C, the larger plastic paper clips she found in stores were "very flimsy and would slip off" the thick packages of material she wished to clip together. Metal binder clips were the standard alternative, but she had had "several bad experiences with being pinched and cut," and so she avoided those types of clips. She therefore set out to design a new clip that not only holds thick piles of assorted papers but also can keep them separated in layers. The development of the clip went through various stages to get the right spring and stiffness to the injection-molded parts, but now that she has perfected it, Suzy Chung is marketing her Layer Clip in various colors and sizes. She holds several other patents, including ones for a chestnut cutter, a hair curling device, and a shovel for weeding, and she uses her Layer Clip to distribute marketing information for all of her products.

An even more recent patent for a paper clip (Fig. 2.13) was issued on July 19, 1994, more than nine years after the application was first filed, to Linda A. Froehlich and Richard D. Froehlich, who own and operate the Ace Wire Spring and Form Company near Pittsburgh. The paper clip patented by the Froehlichs also addresses the problem of clipping large quantities of paper together, and hence the clip they began manufacturing was about three times the size of a standard No. 1 Gem. At first glance, the larger clip might look like just an enlarged Gem, and this may explain why the Froehlichs had to take their case all the way to the Patent Office's appeals board before receiving their patent. In fact, the new clip is not only larger but has much longer end legs than the standard Gem. While George McGill's 1903 patent also had this feature, the Froehlichs' is further distinguished by the nature of the wire from which it is made. Most clips are made of standard steel wire, but the patent for the new clip discloses how a better quality of spring wire can be used to give the clip the ability to hold large amounts of paper securely and yet spring back to its original shape when removed.

The larger amount of better quality wire obviously makes this new clip more expensive to manufacture and thus more expensive to buy, but its advantages over existing large-capacity clips is expected to overcome this disadvantage. Patent drawings (Fig. 2.14) show how the long legs do not dig into the paper the way those of standard clips do, and the Froehlichs believe their four-inch-long clips are also superior in other ways. According to Linda Froehlich, butterfly clamps (also known as Ideal paper clips) do not stay on as well, binder clips (the squarish black clips that are

US005329672A

United States Patent [19]

Froehlich et al.

[11] **Patent Number:** **5,329,672**

[45] **Date of Patent:** **Jul. 19, 1994**

[54] **METAL WIRE PAPER CLIP STRUCTURE**

[76] Inventors: **Linda A. Froehlich; Richard D. Froehlich,** both of 5151 Keiners La., Pittsburgh, Pa. 15205

[21] Appl. No.: **604,970**

[22] Filed: **Oct. 29, 1990**

Related U.S. Application Data

[63] Continuation of Ser. No. 45,452, May 4, 1987, abandoned, which is a continuation of Ser. No. 764,566, Aug. 12, 1985, abandoned.

[51] Int. Cl.⁵ B42F 1/04
[52] U.S. Cl. 24/67.9
[58] Field of Search 24/67.9, 546, 545, 547

[56] **References Cited**

U.S. PATENT DOCUMENTS

742,892	11/1903	McGill	24/547
742,893	11/1903	McGill	24/547 X
2,269,649	1/1942	Comley	24/547
4,017,337	4/1977	Winter et al.	24/547 X
4,597,139	7/1986	Lau	24/546

FOREIGN PATENT DOCUMENTS

665847 9/1938 Fed. Rep. of Germany 24/546

Primary Examiner—James R. Brittain
Attorney, Agent, or Firm—Raymond N. Baker

[57] **ABSTRACT**

Improved paper clip is formed from a single piece of spring-quality metal wire to have an elongated U-shaped inner loop nested within an elongated U-shaped outer loop. Each loop includes a free leg and a connector leg; the longitudinally-extending leg portions of the free leg and connector leg of each respective loop are of substantially the same length. The U-shaped loops are joined together by an arcuately-curved interconnector extending between the connector leg of each loop. Distal ends of the free leg of each loop are located contiguous to such arcuately-curved interconnector which defines one longitudinal end of the paper clip; with such new configurational and other features, damage to outer surfaces of stacked paper by such distal ends is avoided and clasping force is applied along substantially the full length of such free legs during use of the improved clip.

5 Claims, 2 Drawing Sheets

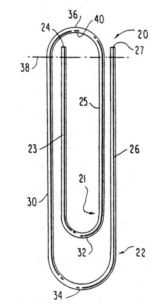

FIGURE 2.13 Patent for a large paper clip made of spring wire, issued to Linda and Richard Froehlich

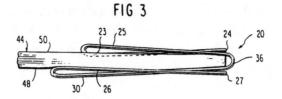

FIG 3

FIG.4 (PRIOR ART)

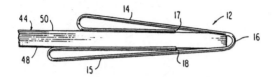

FIG.5

FIG.6 (PRIOR ART)

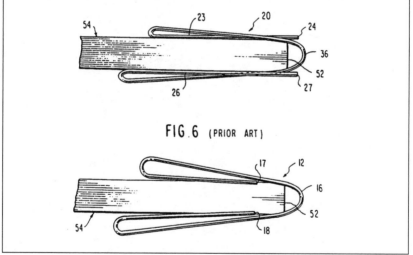

FIGURE 2.14 Patent drawings showing how the ends of conventional paper clips dig into papers, thus tending to tear them upon removal, and how the Froehlichs' clip overcame this problem

opened with fold-down arms) are too bulky, and her clips are easier to put on.

These examples are only some of the many patents for paper clips that have been granted in recent years. It is unlikely that there will be an end to new ideas for paper clips any time soon, since it is unlikely that any single design will satisfy all of the competing objectives that engineers, inventors, and users have long recognized in its design requirements.

3

Who has not experienced the frustration of the lead of a pencil point repeatedly breaking? It appears to happen especially easily in thin-lead mechanical pencils, where more lead seems to end up in the wastebasket than is used up in writing, and at times the inconvenience of the constantly breaking lead seems to negate the advantages of not having to sharpen those familiar 0.5-mm thin-lead pencils. But rather than just complain or ignore this little annoyance of everyday life, let us look more closely at pencil points to see why they break and what they can teach us about materials and their strength and about how engineers analyze problems.

If a pencil point is pushed hard against a writing surface, it in turn pushes equally hard back (according to the principle of action and reaction of forces, which Newton articulated as his Third Law). When the push is more than the material of the pencil point can withstand, the pencil point breaks. To understand this phenomenon in more detail, we can look more closely at the lead protruding from the end of a mechanical pencil.

CANTILEVER BEAMS

A pencil is a structure, because it is an assemblage of materials designed to sustain loads. More specifically, a pencil used in writing or drawing can be described as a cantilever beam, which is simply a structure supported at one end only. Trees blown by the wind, airplane wings with attached engines, and apartment house balconies holding partygoers are other common structures that may be described as cantilever beams. The most important part of a pencil cantilevered from the hand holding it is clearly the pencil point, which the writing surface pushes against as the pencil is used.

Because we recognize that the metal and plastic parts of a mechanical pencil are so much stiffer and stronger than the small piece of lead

FIGURE 3.1 Galileo's illustration of a cantilever beam

projecting out from the end of the metal sleeve, we can imagine that the pencil proper is for all practical purposes an extension of the hand, with the exposed lead itself comprising the cantilever beam. The juncture between the projecting lead and the pencil proper may be considered the root of the beam. Even though the pencil is really at an angle to the paper, and even though the push of the paper is really at an angle to the pencil-beam, there are enough similarities between pencil points, trees, wings, and balconies for engineers to see them as analogous. Thus, understanding the behavior of any one of these examples gives us more insight into beams generally.

FIGURE 3.2 Thin-line polymer lead in a bending test, demonstrating its flexibility

Figure 3.1 shows a cantilever beam as illustrated in Galileo's important book, *Dialogues Concerning Two New Sciences,* published in 1638. One of the two new sciences of the title is today known as strength of materials, among the first courses engineering students take after calculus and physics. It was Galileo's investigations into the behavior of the cantilever beam that laid the foundations for this course, and while much of what he wrote has been superseded, his general approach to the problem remains a model of sound engineering science. What motivated Galileo to look so closely at the cantilever beam was the inexplicable breaking of wooden ships, marble columns, and other products of Renaissance engineering. Broken pencil points can do the same today.

While such diverse structures as ships and pencil leads might not seem at first to be related to each other, let alone to cantilever beams, they are. Figure 3.2, for example, shows a piece of polymer-based pencil lead being tested for its strength and flexibility. In this case, the pencil lead is supported at both its ends, and the load is being applied at the center.

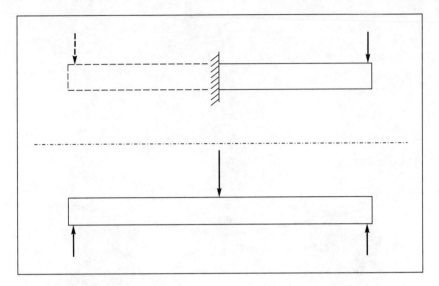

FIGURE 3.3 Demonstration that a simply supported beam (bottom) may be thought of as the mirror image of two back-to-back cantilever beams

This is not unlike the situation in which a ship with a heavy load in its center hold might be found when its bow and stern are raised on high waves. In such an arrangement the ship or lead is considered to be a simply supported beam, but it in turn can also be considered two cantilever beams, as Fig. 3.3 demonstrates. Across the central vertical plane of symmetry, each half of the beam can be considered a wall into which the root of the other half is embedded. The support loads can then be viewed as forces or weights which bend each of the cantilever beams. Because each of the halves is the mirror image of the other, analyzing either one is sufficient for approaching the entire problem. Such different points of view are often helpful in analyzing new engineering problems in terms of older, familiar ones, about which something is already known.

Galileo saw a cantilever beam as a bent (or canted) lever, with its fulcrum at the point labeled B in Fig. 3.1. He imagined that the effect of the boulder hanging from the end of the beam was to produce a tendency to turn the lever about point B, but that such motion was resisted by a uniform force of cohesion across the entire section AB, as shown in Fig. 3.4a. By balancing these opposing effects, Galileo could relate the capacity of the beam to support a weight to the size of the beam's cross section and the strength of the material of which it was made. The conclusions

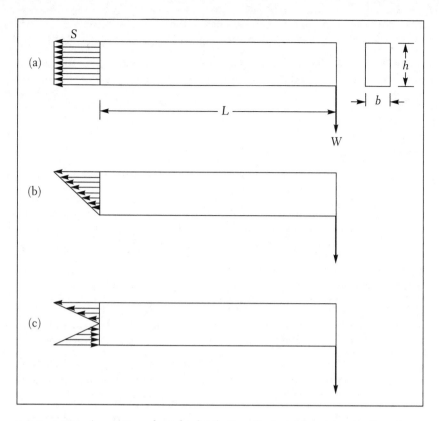

FIGURE 3.4 Assumptions about the distribution of cohesive forces in a cantilever beam by: (a) Galileo, (b) Mariotte, and (c) Parent

Galileo reached made sense to him at the time, but they overestimated the capacity of the beam by a factor of three.

Galileo's assumption that the resisting force was uniformly distributed across the entire section AB did not allow for the fact that timber, like all materials, has a degree of elasticity, and its force of resistance would be proportional to the stretch of the beam, which should increase with increasing distance from the pivot point. Such a realization led subsequent analysts, such as the Frenchman Edmé Mariotte, to modify the distribution of resisting force to the triangular form shown in Fig. 3.4b, and this gave a more accurate result. However, this was still an incorrect analysis of the forces keeping the beam in equilibrium, because neither assumption provides a force to balance the net horizontal force across the section AB. In time, still other analysts, such as Mariotte's compatriot A.

Parent, realized that the true action of the forces acting across section AB of the beam is as shown in Fig. 3.4c, with the bottom being pushed into the wall and the top being pulled out. There are also additional forces acting on the beam, and their careful and systematic analysis is the subject of study in the engineering science courses known as statics and strength of materials.

MECHANICAL PENCIL POINTS

In Fig. 3.5 only the germane part of a mechanical pencil point is shown, with the force acting vertically upward, as it would on a very slick writing surface (one with virtually no friction between the pencil and paper) or a piece of glass. We can imagine that this force has two components, which give two different effects. One component, which is along the axis of the pencil lead, wants to push the lead back into the pencil's sleeve, as it does when we release the gripping mechanism to retract the lead. We also experience the effects of such a force when the lead inside has gotten so small that it does not reach back to the gripping device. (In this case, we propel the last half-inch or so of lead out of the pencil, discard it, and maneuver a fresh long piece of lead in place to resume writing, thus demonstrating one definite shortcoming of mechanical pencils that cries out for improvement.) The other component of the push of the paper on the lead acts perpendicular to the axis of the lead and wants to bend the lead (as in Fig. 3.2), and when this bending is more than the material of the lead can take, the point breaks off. (It is easy to see how bending breaks the lead by experimenting with a long piece of lead before it is inserted into the pencil. If you prefer not to use pencil lead, pieces of dry thin spaghetti or vermicelli will also demonstrate the phenomenon under discussion.)

Since the part of the paper's push that drives the lead back into the pencil does not appreciably affect the bending action, we can ignore it when looking into the breaking of the lead. In this case, the only relevant action on the end of the pencil point is one that is perpendicular to the shaft of lead. Furthermore, whether this is imagined as a push from one side or a pull from the other makes little difference, for both actions bend the lead equally. Hence, we can represent the pencil-lead cantilever beam simply as shown in Fig. 3.6a, which makes it completely analogous to Galileo's problem of the beam projecting from a wall and supporting a heavy boulder at its end.

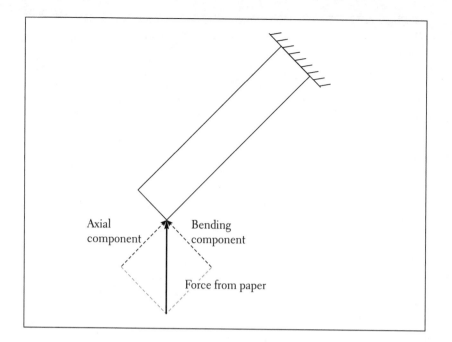

FIGURE 3.5 Idealization and decomposition of forces acting on a mechanical pencil point

Just as Galileo observed that a cantilever beam (without any notches, knots, nicks, or other anomalies that complicate the problem) will break at the wall, so we can observe that the lead comprising a mechanical pencil point (without any nicks, bad spots, or other anomalies) will invariably break at the juncture with the metal or plastic sleeve. To see that the parenthetical caveats are not just theoretical diversions, use a pocket knife or scissor blade to saw part way through a piece of lead and propel the lead so that the nick is located some distance from the metal sleeve. With the nick facing downward, as shown in Fig. 3.6b, press the pencil point onto a writing surface until the lead breaks. If the nick is deep enough, the lead will break at the nick and not at the juncture with the sleeve.

With no nick, the mechanical pencil lead breaks at the sleeve, just as Galileo's beam breaks at the wall, because that is where the intensity of force, which grows larger toward the root of the beam, first exceeds the strength of the material. The lead-beam can break at some distance from the wall if a nick or notch raises the intensity of force the beam must carry at that location over what its strength is there. Clearly, the presence of

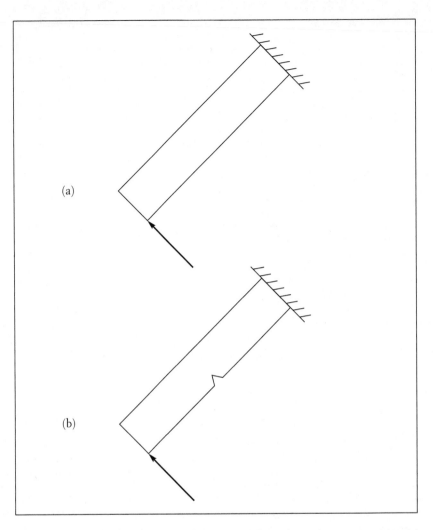

FIGURE 3.6 Further idealization of force on a mechanical pencil point comprising: (a) perfect lead, and (b) notched or nicked lead

the nick changes the behavior of the pencil-lead beam as readily as a saw cut could the behavior of Galileo's beam or the limb of a tree. Such details complicate engineering analysis, but do not make it impossible.

Thin-lead mechanical pencils followed the development of polymer-based leads, which are very flexible (that is, capable of sustaining a fair amount of bending, as shown in Fig. 3.2) and, in spite of how easily these leads seem to break under writing conditions, are relatively strong for their size. Before the pencils holding such leads became so popular, the standard mechanical pencil lead was a ceramic rod (made principally of graphite and clay) over a millimeter in diameter, which made it relatively thick and thus not capable of drawing so fine a line as the newer plastic leads. Ceramic pencil leads had to have larger diameters because they were so brittle and would break too easily if thinner.

There are clearly two competing objectives in pencil making: (1) to make the lead as thin as possible, so that as fine a line as desired can be drawn, and (2) to make the lead thick enough to give it sufficient strength. In pencil making as in most engineering endeavors, in addition to often conflicting goals there are always additional objectives which further complicate the problem. A pencil must be of a comfortable size, so that it can be used for writing over an extended period of time. To make a pencil lead itself (with no case) so thick as to be comfortable in the hand defeats the objective of achieving a relatively fine line, and also, since the writing substance is the single most expensive ingredient in the pencil, a comfortably thick lead would be wasteful because so much of it would be lost in sharpening. Furthermore, the very graphite in pencil lead (a misnomer) that makes it write also dirties the hand that holds it, and so an unprotected piece of lead is not very desirable anyway. For these reasons and more, the wood-cased pencil evolved shortly after the discovery of graphite.

WOOD-CASED PENCILS

A wood-cased pencil sharpened to a conical point presents a more complicated problem in analysis than does a mechanical pencil because of the more complex geometry and the interaction between the wood and the lead. Up until the 1930s the points on wood-cased pencils generally broke in a rather messy way: The lead most often snapped inside the wooden shaft and splintered the wood back in the process. This mode of behavior was likely to occur for two main reasons: (1) If the pencil happened to be badly warped or dropped on a hard surface, the lead inside the wooden case became broken into several short pieces, which in effect meant that the pencil point was not so deeply embedded in the wood and so was a poorly supported cantilever beam. (2) When the pencil point was sharpened, the wood was necessarily thinned in the very place where its strength was needed to support the projecting lead. In both cases, the action of the pencil point as a cantilever beam was compromised at the critical location—which is analogous to stones being removed from the wall above Galileo's beam. If only a few wall stones remained above the beam support, they could easily be displaced by the lever action of the beam, which would then collapse. Similarly, the thin sheath of wood near the sharpened pencil point can provide little support to the lead, especially if there is a poor bonding between the lead and wood.

Such undesirable features of wooden pencils were the focus of study and analysis by engineers in the 1930s, and understanding the shortcomings was the goal of engineering efforts that started with measurements to determine the comparative strengths of competing pencil brands. Thus, when the Eagle Pencil Company wanted to advertise its Mikado pencils as stronger than those of the opposition, pencil points were broken under controlled conditions. As it turned out, the Eagle pencils were not so significantly stronger that the company could support the desired advertising claims. Hence, engineers were asked to see if they could make Eagle pencils stronger. Just as Galileo did three centuries earlier for timber beams and marble columns, the twentieth-century engineers began by looking more closely at exactly how pencils did break, and analyzing the phenomena involved. They soon realized that the problem lay principally in the fact that the wood did not act adequately in concert with the lead to help prevent its breaking prematurely in the case or its tearing through the hollow cone of wood from which the sharpened lead projected. Understanding this weakness of the lead-wood composite beam enabled the engineers to focus on how to strengthen it. In general, it is the analysis of failures or of limitations of a design that leads to improved understanding and thus to improved products.

Wood-cased pencils were made by gluing rods of pencil lead into grooved wood slats and then gluing on a mating piece of wood, as shown in Fig. 3.7. When the glue was dry, the pencils were shaped, finished, packaged, and shipped from the factory. Unfortunately, the glue did not always form a very good bond between the lead and the wood, for various reasons. Among these were that both the lead and the wood were impregnated with wax—the lead to make it write more smoothly and the wood to make it easier to sharpen. The presence of the wax prevented the glue from adhering as well as it might, and as a result the lead and wood did not bond tightly together. (It is possible to find older pencils that look perfectly fine but in which the wood and lead are completely unattached, and the lead can be slid freely back and forth inside the wooden shaft.) When a poorly bonded pencil was dropped, the lead could easily break into many smaller pieces inside the wood case, often to be discovered only when the pencil was sharpened. A common frustration was to have

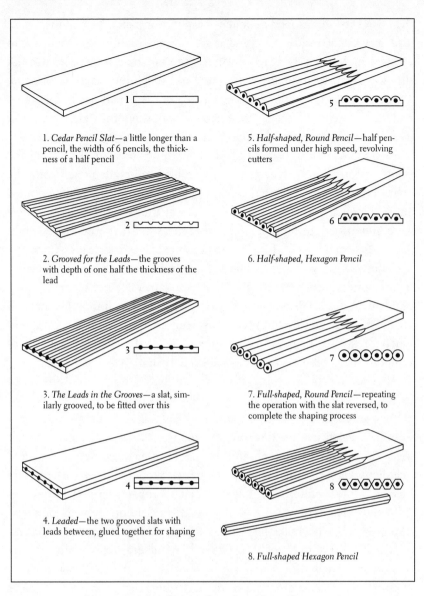

1. *Cedar Pencil Slat*—a little longer than a pencil, the width of 6 pencils, the thickness of a half pencil

5. *Half-shaped, Round Pencil*—half pencils formed under high speed, revolving cutters

2. *Grooved for the Leads*—the grooves with depth of one half the thickness of the lead

6. *Half-shaped, Hexagon Pencil*

3. *The Leads in the Grooves*—a slat, similarly grooved, to be fitted over this

7. *Full-shaped, Round Pencil*—repeating the operation with the slat reversed, to complete the shaping process

4. *Leaded*—the two grooved slats with leads between, glued together for shaping

8. *Full-shaped Hexagon Pencil*

FIGURE 3.7 Steps in the manufacture of wood-cased pencils

the point drop out as soon as pencil was put near paper. Even if the pencil lead was intact inside the wood case, excessive writing pressure could create a high intensity of force where the sharpened wood case was not strong enough to resist it (what pencil manufacturers referred to as the "pressure point"), and the wood on the top of the pencil would split open and the exposed lead break. It was as if the wall supporting Galileo's beam were made of loosely held stones that were pried up by the cantilever.

Engineers learn a lot by studying how things break and fail, and the troubleshooting and research into the problem of fragile pencil points by engineers at the Eagle Pencil Company led to this understanding of how and why pencil points were breaking the way they were. Once the causes were understood, further research led to measures that could be taken to lessen the occurrence of failure in wood-cased pencil leads. Since the presence of wax was responsible for the inability of the glue to form a secure bond, means were sought for negating the effects of the wax. These problems did not fall neatly into purely chemical or purely mechanical engineering categories, but such cross-disciplinary research is common to much of the work of the engineer and is what makes the pursuit of solutions to problems always fresh and challenging.

After some experimenting, it was found that bathing unsheathed lead in sulfuric acid burned off the wax from the surface, and a subsequent bathing in calcium chloride coated the lead with a sealing film of gypsum. The wood in turn was impregnated with a resinous binder, which made the wood fibers tough and less easily split. In combination, these treatments of the components of the pencil prepared them for glue that held as never before. The Eagle Pencil Company termed its process "Chemi-Sealed," and other manufacturers soon countered with their own improved ways of making pencils, labeling them "Bonded," "Wood-clinched," "Pressure Proofed," and the like. Advertising departments lost no time in bringing the advantages of the new pencils to the attention of consumers. The trademarked terms describing how pencil leads were more fully supported by the wood casing still appear on pencils today, and they are indications that the cantilever beam of lead is held as securely in the pencil shaft as Galileo's beam was assumed to be held in the masonry wall.

BROKEN-OFF PENCIL POINTS

The points of the improved pencils did not break in the same way, of course, but that is not to say they did not break. Which of us has not experienced the frustration of sharpening a pencil to a fine point just to have it snap off as soon as we touch it to paper? Such experiences teach us to press more lightly on a newly sharpened pencil, but we can easily forget in the midst of writing something that absorbs all of our mental energies, and we can thus find our pencil points breaking with annoying frequency. If we are really absorbed in our writing task, we will almost unthinkingly resharpen our pencil, pick up a fresh one, or continue to write with the increasingly blunting (and actually strengthening) pencil in our hand.

Don Cronquist, a California engineer, did just such things while preparing a long hand-written rough draft of a report that he was working on in the mid-1970s. We can imagine that his desk was piled high with books, papers, notes, and so forth, much as ours would be after working on an important term paper, laboratory report, or speech. Whether or not we would clear our desk before going on to another project, Cronquist proceeded to clear his and "became mystified" when he "discovered a very large number of broken-off pencil points (BOPPs) lying between and behind the books and other reference materials" on his desk. It was not the large number of pencil points that was remarkable, for he knew how easily they broke off newly sharpened pencils. Rather, according to Cronquist, "the mysterious thing about the collection of BOPP's was that they were almost all nearly identical in size and shape."

An observation like Cronquist's, that all the BOPPs looked similar, suggests to an engineer or scientist that there is an engineering or scientific explanation for how the pencil points broke the way they did. The BOPPs lying about the desk were literally data points from an experiment, albeit an unplanned one, and their falling into the strong pattern of similarity that Cronquist noticed presented too neat a problem in analysis to pass up. Just as scientists seek to understand the behavior of the given natural world, whether it be the growth of plants or the motion of planets, so engineers seek to understand the behavior of objects crafted, fabricated, or manufactured by humans. Such objects are known as artifacts, and when their nature and behavior are the subject of study by engineers, the engineers are said to be engaged in analysis or engineering science. Galileo laid the foundations of engineering science when he attacked the problem of the cantilever beam the way he did.

Since Cronquist saw no obvious explanation for the size and shape of BOPPs in the geometry of how his pencils were sharpened or in the material of the pencil lead itself, he did what Galileo did three centuries earlier—Cronquist proceeded to attack the problem with an analysis that combined geometrical and strength considerations. Expecting that the intensity of force (known as stress) in a tapered cantilever beam (the pencil point) depended upon distance from the load or force, Cronquist argued that a pencil point would break at the location where the stress in the lead reached a limiting value, which is the breaking strength of the material. He assumed that this strength (or ability to resist stress) was the same everywhere in the pencil lead, because his were good quality pencils that he could assume were made of good raw materials and manufactured to tight standards. He further assumed that the geometry of the pencil point was a regular truncated cone, so that he could apply mathematics no more complicated than geometry, algebra, and calculus and carry out calculations with pencil and paper alone. The advantages of this approach over resorting right away to a computer model of the problem are that one can solve once and for all the general problem explicitly and get answers in terms of significant parameters. A computer program could then be used to advantage to explore myriad different configurations of pencil point angle, writing angle, and so forth. (The disadvantage of using a computer initially and directly for such a problem, and for much more complicated problems also, is that only one specific combination of geometry, loading condition, and material strength can be considered at a time. An understanding of general principles comes only if one can infer them from the collection of separate solutions.)

The principal complicating factor in Cronquist's analysis of the pencil point is that it is tapered rather than cylindrical. This means, of course, that there is not a constant area to resist the load over which the breaking force is distributed. Since Galileo's beam has constant area throughout, it is a simple matter to conclude, as Galileo did, that breaking will occur where the intensity of breaking force is greatest, which by the principle of the lever is clearly at the wall. In the case of the tapered beam of the pencil point, the area increases with the diameter squared, and so there is no obvious breaking location. A general analysis must be carried out to calculate where the intensity of load first exceeds strength.

Cronquist began his analysis by modeling the pencil point as a cantilever beam with a tapered circular cross section and assuming that the force of the paper on the pencil is perpendicular to its axis, as shown in Fig. 3.8. (While this assumption about the force is not likely to be realistic in most writing situations, it is a reasonable one with which to begin an engineering analysis because, just as with the mechanical pencil point, it

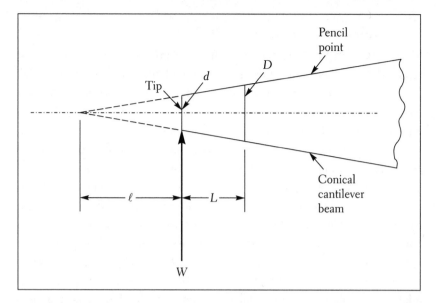

FIGURE 3.8 Engineer Don Cronquist's idealization of a wood-cased pencil point as a cantilever beam

is most likely the component of the force perpendicular to the pencil axis that has the most influence on bending the point until the strength of the lead is reached and a broken-off pencil point results. This can be verified with a piece of chalk by comparing how much more easily it is broken by a push perpendicular to the axis of the chalk as opposed to a push or pull directed along the length of the stick of chalk.) Once he had defined the assumptions about the applied force and geometry of the problem with the aid of a diagram similar to the one in Fig. 3.8, Cronquist was able to begin his more mathematical analysis. Developments since Galileo's time have made the problem of calculating the bending stresses in a cantilever something that second-year engineering students are expected to master. After some calculation, Cronquist did indeed find his equations to predict that the pencil point should break at some intermediate location between the writing tip and the juncture with the wood case. Cronquist had a degree of confidence in his results in that they correctly predicted that sharper pencil points break more easily than blunt ones, which was consistent with his experience. Furthermore, the size of the BOPP that his equations predicted matched very closely the size of the BOPPs he had found on his writing desk.

Jearl Walker, in his "Amateur Scientist" column in *Scientific American*, reported on Cronquist's study and presented further experimental evidence that his simple analysis was sound. However, like Cronquist, Walker could not provide a satisfactory explanation as to why the BOPPs had fracture surfaces that were distinctly slanted back toward the pencil body. Rather than breaking across planes perpendicular to the edge of the pencil point the way a piece of chalk tends to, all the BOPPs had clearly formed by breaking off at a slight angle, as illustrated in Fig. 3.9. The consistency of this angle suggested that there should be some definite geometrical and/or strength-of-materials explanation for the phenomenon, but the fact that neither Cronquist nor those who scrutinized his experimental data and analysis could come up with a satisfactory explanation left not a small measure of mystery associated with what had seemed to be a relatively simple problem in analysis with a rather straightforward analytical solution. (Some readers of Walker's column in *Scientific American* subsequently offered an explanation of the phenomenon, which involved taking into account forces oriented across the width of the pencil lead as well as along its length. Such additional forces, known as shear forces, are among the complications that engineering students learn to analyze in a strength of materials course.)

Cronquist was not able to explain the slanted fracture surface on his BOPPs for reasons analogous to those that kept Galileo from catching an error in his seminal analysis of the cantilever beam. Both Galileo and Cronquist made some elementary assumptions at the outset about how their beams would break under certain realistic but simplified forces, and they proceeded to interpret their subsequent analysis in the context of their assumptions, continuing to take as a given what they had assumed at the outset. Galileo not only assumed that the cantilever beam would fail at the wall, which is correct, but that the breaking force would be uniformly distributed across the plane that would fracture, which is incorrect. Cronquist (with three centuries of inherited experience) did not assume so much. He allowed the force in the pencil lead to vary across the planes perpendicular to the axis as well as in intensity along the axis of the pencil-beam, but he ignored any effect of other forces when he assumed that fracture would occur where the maximum intensity of the axial force reached the strength of the pencil lead. Galileo had checked his own result by noting that it correctly predicted that a deep beam was

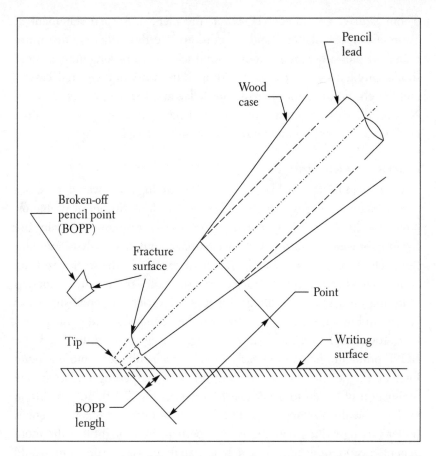

FIGURE 3.9 Adaptation of Cronquist's drawing demonstrating details of how a pencil point breaks at an angle to the pencil axis

stronger than a shallow one, and this gave him confidence that he had done his analysis correctly, even though it was no conclusive proof of correctness. Over three centuries later, Cronquist checked his result against the BOPPs that had prompted his analytical investigations, and when the predictions of his result and their dimensions agreed, he took that as confirmation of his starting hypothesis, even though he could not explain totally the slanted fracture surface. It would take subsequent and fresh analyses of the pencil point problem to find the error.

While our engineering, science, and technology have clearly advanced a great deal over the centuries, the *method* of how we analyze and attack problems remains essentially unchanged since Galileo's work. Even when we use the most sophisticated computer models available, the same errors of omission can be made. However, the clear and open statement of assumptions, calculations, and conclusions like those that Galileo introduced enables subsequent engineers and scientists attacking the same and similar problems to check and build upon the work of their predecessors and thereby advance the state of knowledge and the state of the art. Isaac Newton likened this cumulative effect of being able to see further than one's predecessors to standing on the shoulders of giants.

FURTHER ANALYSIS

As the case of BOPPs makes clear, it is the nagging unanswered questions that so often drive engineers and scientists to look further at problems. Not only did Cronquist's analysis leave open the question of why the fracture proceeded across a slanted surface, but his analysis was also limited to the case of a certain kind of force exerted by the writing surface. It may have been simply the desire to carry out a more general analysis (allowing for a rougher writing surface, for a variety of pencil sharpener angles, and for different angles at which the pencil was held to the paper) that drove Stephen Cowin, another engineering scientist, to attack the BOPP problem anew in the early 1980s. By allowing for a more general force on the pencil point, an arbitrary angle of sharpening, and a variable inclination of pencil to paper (as suggested in Fig. 3.10), he was able to explore how the size and shape of BOPPs might vary for different pencils under varying writing conditions. And perhaps he thought that the more general assumptions under which his analysis was carried out would provide insight into why the slanted surface occurred.

As it turned out, for all of his generalizing the geometry and loading conditions of the problem, Cowin did not question the more fundamental assumption of the criterion for fracture, and he also assumed that failure would occur where the greatest value of axial stress reached the tensile strength of the material and also tacitly assumed that, once started, the fracture would proceed in a straight path across the pencil point. Thus, like Cronquist, Cowin essentially assumed that fracture would result in a BOPP that was a regular or a negligibly skewed cone rather than the distinctly skewed one that actually results when a pencil point breaks.

In summary, neither Cronquist nor Cowin could fully explain why the fracture of pencil points occurs across a distinctly skewed plane because they based their entire analysis on the *assumption* that fracture occurred when the maximum tensile stress reached a critical value across a particular plane and they ignored altogether any shear forces or stresses that might

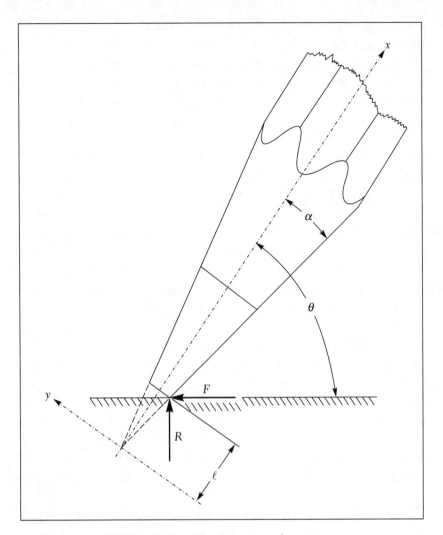

FIGURE 3.10 The geometry of, and forces on, a pencil point

act parallel to that plane and serve to divert the propagating fracture from its initial path. The fact that early attempts at solving a problem like the broken pencil point failed in no way diminishes the contributions of analysts like Cronquist or Cowin in advancing our knowledge and ability to attack the problem anew and with the clear advantages of the foundations they laid. Indeed, the work of such analysts in defining the problem and setting down their approaches to its solution guide and benefit all subsequent analysts, just as Galileo's fundamental but incomplete work on the cantilever beam in his *Dialogues Concerning Two New Sciences* laid the foundation for all modern calculations in problems relating to the strength of materials.

When calculations proceed from assumptions to predict something so accurately as the relative strength of cantilever beams or the absolute size of BOPPs, it also becomes increasingly unlikely that the validity or completeness of the fundamental assumptions are questioned. But just because those assumptions were not invalidated by results, it does not follow that the assumptions were complete or even correct. What determines the location and *initial* orientation of the fracture surface is indeed the greatest stress that the brittle pencil lead can withstand. In fact, while it is the plane across which the *absolutely* greatest tensile stress acts that determines where and how the fracture of the BOPP begins, as the fracture proceeds, the shear component of stress can play an increasing role in determining the plane across which the final separation of the point occurs, as was discussed in the *Scientific American* column that followed up on the original treatment of the problem.

In the geometry of Galileo's beam and in the rectangular geometry of most cantilevers, the greatest intensity of force pulling the material apart (the maximum tensile stress) does act parallel to the long axis of the beam at its outer fibers. However, the conical geometry of a pencil point complicates the problem, and ignoring even so seemingly small a detail limits the efficacy of all subsequent analysis. Whenever there is a boundary surface that has no force applied directly to it, like the sides of a pencil point, then the maximum tensile stress must always act across a plane that is perpendicular to the free surface. In the case of the pencil point, this observation leads immediately to an explanation as to why the BOPPs begin to break off at a slant; and a reconsideration of the basic assumption of the nature of the forces involved, including the consideration of shear, leads to a clarification of how the fracture proceeds.

ANALYZING ANALYSIS

The problem of BOPPs illustrates how important it is to keep in mind and to go back to fundamental assumptions in scrutinizing and interpreting the results of engineering analysis. As long as one does not question the validity or recognize the restrictiveness of basic assumptions, one can overlook the fact that they are limiting one's interpretation of results. As relatively simple and inconsequential as the problem of BOPPs may appear to be, such a lesson is invaluable for developing a keen critical sense of engineering method and judgment. Analysis carried out with the aid of the most complex of computer models is no less subject to such myopia. Indeed, the more complex the problem and its analysis, the more readily we can fall into the trap of focusing so intently and narrowly on the results of the analysis that we forget the fundamental assumptions upon which the analysis was based. Furthermore, as such analysis evolves to deal with increasingly more general or larger systems, whether they be BOPPs or suspension bridges, the more danger there is of losing sight of how restricted by fundamental assumptions the analysis of the model really is.

Everyone makes mistakes, even geniuses like Galileo. Engineers must always be alert for what they may be oversimplifying and overlooking or to what conclusions they may be jumping. Because errors in engineering can have disastrous consequences, it is especially important for engineers to be reflective and alert in their design and analysis. One way of being better prepared for catching one's own errors is to be familiar with the kinds of mistakes that have been made by others, whether three centuries ago or yesterday. We can gain further insight into the kinds of pitfalls there are in engineering design and analysis by understanding the similarities in the way in which even someone so brilliant as Galileo could be misled by his oversimplified analysis of the cantilever beam, in which he ignored variations in the resisting force, and the ways in which we can overlook critical details in the analysis of such things as the fracture of pencil points.

Even actual destructive experiments might not have uncovered Galileo's error easily because real materials and structures seldom occur in conditions as pure as mathematics. Real materials have variations and imperfections that make strength meaningful only in some statistical sense. If we break several objects, ostensibly identical, we are likely to get a range of strengths because the specimens will have varying kinds of imperfections (knots, cracks, splits, voids), and the experiments themselves will suffer from imperfect measurement, lack of alignment, and so forth. Furthermore, even if the results were scattered about a different value than that predicted, the difference might be attributed to something like an inability to get a true cantilever support rather than the real cause. (In Galileo's case, his overlooking entirely the springiness of the material obviously prevented him later from incorporating its important effects into his analysis.)

Because engineers know that theoretical calculations and predictions can seldom capture all the variability and detail that exists in reality, they are not expected to be perfect. The way engineers have long used design

Graphical representations of a pencil—a most familiar object rich in symbolism and significance—are commonly found in newspaper cartoons and advertisements, especially at back-to-school time. Perhaps because these depictions are so common, we tend not to look very carefully at them or to note how often they represent objects impossible to find in real life.

Among the most common errors is the portrayal of how a pencil is sharpened. In reality, the ridges of a hexagonal pencil are cut away by a pencil sharpener or knife. Why is it that sharpened pencils are so frequently drawn with the ridges intact? Does artistic license account for this phenomenon, or is there some other explanation?

formulas derived from analysis is to apply them conservatively in order to take into account a multitude of uncertainties. For example, who is to say how carefully masons might build a stone wall around a cantilever such as shown in Galileo's illustration. Furthermore, who is to say how careful the workmen might be in attaching the heavy weight to the end of the completed cantilever? How many of them will stand on the beam during the construction process, thus adding considerable weight beyond that of the boulder being hung from the end? Will they place it gently on the hook, or will they drop it with such force that it will endanger the beam the way a sudden blow does a pane of glass? And what about the weight of the beam itself? While Galileo discussed how it might be taken into account, the formulas we have been discussing do not consider it. To cover the many contingencies that might overload an understrength beam, engineers often divide the predicted weight it can support by as much as a factor of six or eight. Such conservatisms are known as factors of safety, and while they result in structures that are overdesigned, they also result in structures that are more reliable. Though writing with a pencil may not seem to involve such critical and structurally abusive loads as erecting a beam in a wall or subjecting an airplane wing to a storm's turbulence, seeing the similarities between how such apparently diverse structures work or can fail to work is essential for successful engineering.

4

One of the many time-consuming and often frustrating tasks faced daily in the nineteenth century was the fastening and unfastening of the many buttons or hooks and eyes found on articles of clothing, including high-button shoes. With so many fasteners spaced so closely together, it was not uncommon for someone dressing quickly or inattentively to skip a button or a hook, only to find an extra button hole or eyelet at the bottom of a vest or the top of a blouse, requiring a lot of undoing back to the mistake and then redoing. Among the many people who must have noted and even cursed this and other problems with buttons and hooks and eyes was Elias Howe, Jr., the inventor of the sewing machine. Rather than just complain about the problem, however, Howe came up with "certain new and useful Improvements in Fastenings for Garments, Ladies' Boots, and other articles to which they may be applicable," and he was awarded a patent in 1851. His patent consists of one page of drawings (Fig. 4.1) and one page of text.

Howe's device, like all inventions, addressed shortcomings associated with the existing way of doing things, as he clearly stated: "The advantage of this manner of fastening garments, &c., consists in the ease and quickness with which they can be opened or closed, while there is no liability of their getting out of order." The manner in which the new fastening device was intended to function can readily be seen in the patent drawings, and it clearly could work in principle. Certain difficulties in its continued smooth operation can also be easily imagined, however. For example, Howe's fastening device would require that its metal clasps fit snugly, but not too tightly, around the beaded fabric along which they would slide. Assuming that this close tolerance between the clasps and beading could be achieved in manufacture, it is questionable that it could be maintained long in use. As the fastening device was sat upon or bumped while moving about, the clasps would no doubt be bent to the

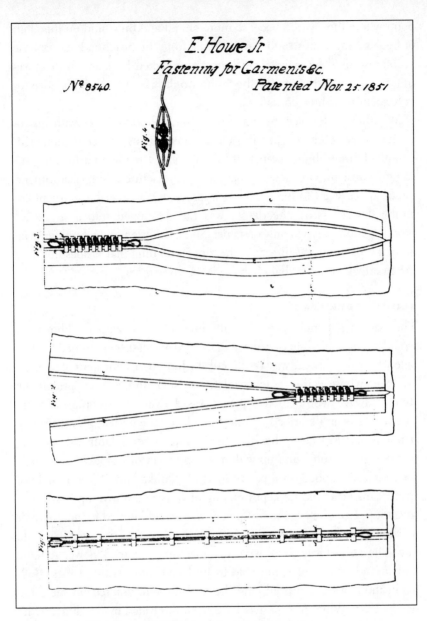

FIGURE 4.1 Elias Howe, Jr.'s 1851 patent for a fastening device

point where they would become more closed and thus bind on the cloth or become more loose and thus pull free from the beading. Even if those problems could be avoided, in time the repeated back-and-forth movement of the metal clasps on the fabric would fray it to an ineffective, or at least an unsightly, condition.

Whether he foresaw these as insurmountable difficulties with his invention or whether he put it aside because of his preoccupation with potentially more lucrative patent-infringement lawsuits that he was pursuing against thriving sewing machine manufacturers like Isaac Singer, Howe appears not to have tried to improve and market his garment-fastening device. Thus, the design was not developed into a successful product, and it survives only on paper. Some historians of technology even deny it a place in the history of the zipper, arguing that it does not have the interlocking teeth that characterize a true zipper.

SLIDE FASTENERS

The person generally credited with inventing the zipper, although it would not be called that for more than 30 years after he obtained his first patents for the device, was Whitcomb L. Judson, a Chicago mechanical engineer whose earlier patents related to such things as a "pneumatic street railway," whose motive power was derived from compressed air. Judson has been described as a portly individual who had grown tired of bending over to lace up his high boots. Thus finding fault with existing technology, Judson came up with a "clasp locker or unlocker for shoes," for which he applied for a patent in 1891. Unlike Howe, Judson did not neglect his idea, and he kept thinking of ways to improve upon his own invention. Even before the first patent was issued he filed for another, for a "shoe fastening" device (Fig. 4.2). Unlike his first idea, which would have altered the way shoes were manufactured, Judson's newer scheme had the advantage of being able to be laced into existing shoes. Both applications were approved, and the patents were granted on the same day in 1893. While Judson may have been motivated by the difficulty he encountered in bending over to tie his shoes, he clearly recognized that his invention had much wider application: "The invention was especially designed, for use as a shoe-fastener; but is capable of general application wherever clasps consisting of interlocking parts may be applied, as for example, to mail bags, belts, and the closing of seams uniting flexible bodies." All these applications would indeed be made, but first the basic

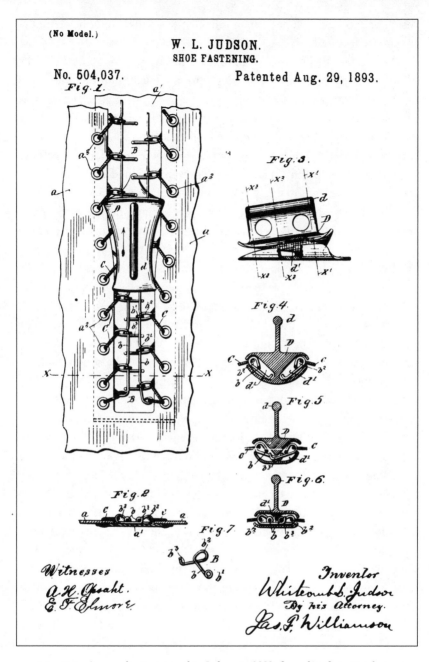

FIGURE 4.2 A second patent issued to Judson in 1893, for a shoe-fastening device

design had to be developed into a reliable device that could be economically manufactured.

The promise of Judson's idea attracted the support of Harry L. Earle, whom Judson had known since they were both salesmen of agricultural machinery and who had been a promoter of the pneumatic railway scheme. Principal financial backing came from Lewis A. Walker, a Pennsylvania lawyer who foresaw a fortune to be made from Judson's invention, and the Universal Fastener Company was formed to exploit the patents. Judson's next two patents, issued in 1896, were assigned to the company, and they show designs that look more substantial. Early Universal fasteners did not sell very well, however, in large part because they tended to pop open at inopportune times and because their sharp edges and pointed ends tended to tear the fabric of what they were supposed to fasten. Furthermore, unless the labor-intensive methods of manufacturing early versions of the devices were effectively automated, the prices of the fasteners could not be kept low enough to make them attractive to potential customers. Thus, Judson had to continue to develop his device while at the same time design a machine to manufacture it.

A decade after his first fastener-patent application, Judson applied for a patent for a "chain-making machine," which was granted in 1902 (Fig. 4.3). This machine was designed to make the "interlocking chains . . . of hooks and links" that were crucial components of a successful fastener. Compared to the earlier patents for the fasteners, this one for the machine is long, with eight pages of drawings and nine of text. This should not be surprising, however, for machines that can automate the manufacture of complex products are even more complex, and can contain a great deal more moving parts, than their products. Unfortunately, neither Judson's machine nor the variety of fasteners that it made was reliable or effective enough. He thus developed a new fastener device, in which the troublesome chains were replaced with hooks and eyes fastened directly to lengths of fabric that could be attached to shoes, garments, and other items, and it was possible to make a simpler machine. In the meantime, the Universal Fastener Company had evolved into the Fastener Manufacturing and Machine Company, which in turn became the Automatic Hook and Eye Company.

The new fastener was marketed under the name C-curity, which clearly implied that it did not share its predecessors' characteristic of popping open when it was not supposed to. Advertisements for the C-curity fastener trumpeted its advantages: "A pull and it's done! No more open skirts . . .

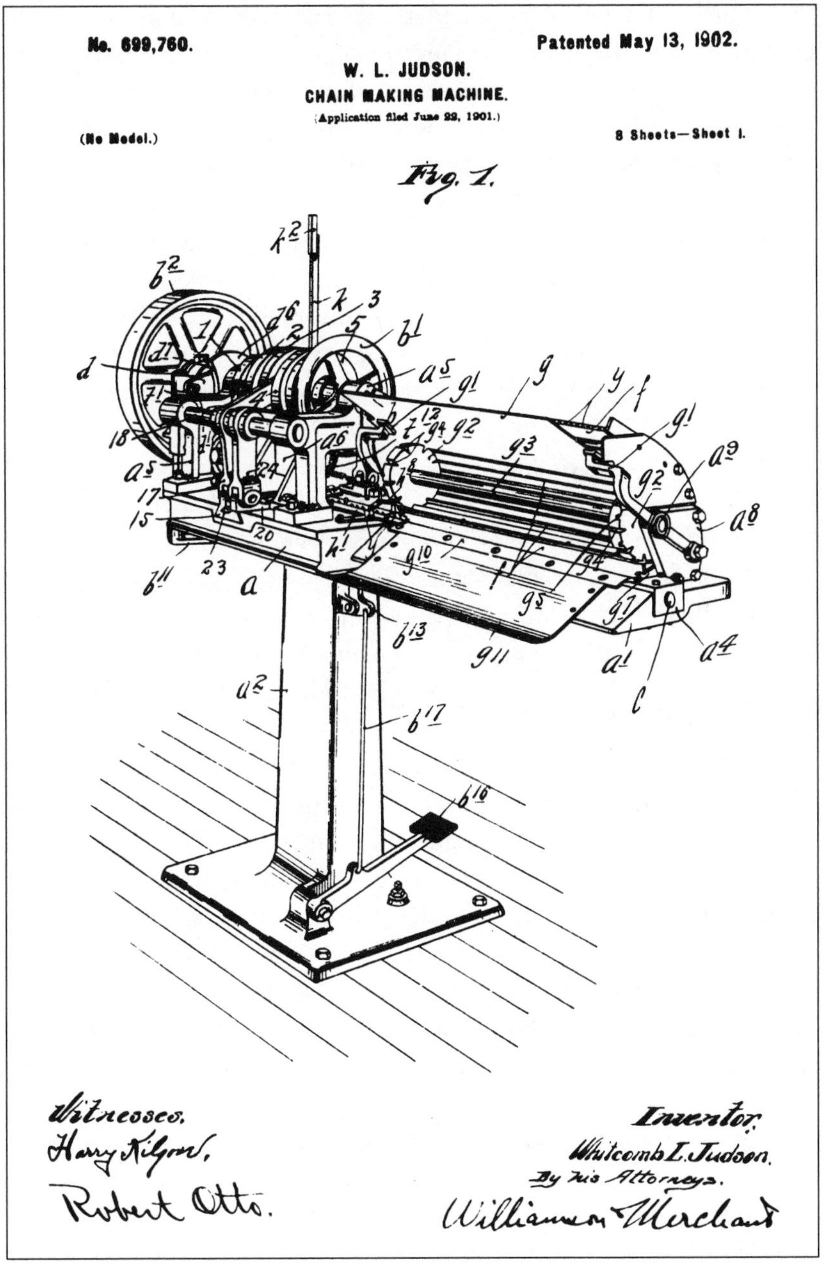
No. 699,760.
Patented May 13, 1902.
W. L. JUDSON.
CHAIN MAKING MACHINE.
(Application filed June 22, 1901.)
(No Model.)
8 Sheets—Sheet 1.
Fig. 1.
Witnesses.
Harry Kilgore,
Robert Otto.
Inventor:
Whitcomb L. Judson.
By his Attorneys.
Williamson & Merchant

FIGURE 4.3 Patent for a chain-making machine

Your skirt is always securely and neatly fastened." Unfortunately, the product did not live up to its promise, and C-curity fasteners were famous for pulling apart when they were supposed to be holding securely together and for the slider getting stuck at the end, locking the embarrassed wearer into an open skirt or pair of trousers. Every manufacturer should want to know of such problems with its products, of course, so that they may be addressed in further development. However, in the case of C-curity, the already wordy and complicated instructions for applying the fastener to garments seemed to convey a lack of uncertainty on the part of the manufacturer: "Customers will confer a favor on us by reporting any difficulty in applying fastener, in which case we will send more detailed instructions."

The Automatic Hook and Eye Company was becoming concerned that Judson's earliest patents were soon to expire, and other inventors were beginning to patent newer versions of what were coming to be known as slide fasteners. Among these was Ida Josephine Calhoun, of Tampa, Florida, whose 1908 patent represented an "improvement in the application of the slide fastener to dresses." At about the same time, inventors in Europe were also being issued patents for slide fasteners. One design that was very similar to what would eventually become the familiar zipper was invented by Katherina Kuhn-Moos and Henri Forster of Zurich, Switzerland, who received Swiss, German, and British patents in 1912.

In the meantime, the Automatic Hook and Eye Company had hired Gideon Sundback, who was born in Sweden and educated in Sweden and Germany as an electrical engineer. He came to America in 1905 and began working for the Westinghouse Electric and Manufacturing Company in Pittsburgh, but a year later went to work for Automatic Hook and Eye as a draftsman and design engineer responsible for further development of the machinery. Sundback was brought to Automatic Hook and Eye by Peter Aronson, who had been responsible for keeping "Judson's machine running long enough and steadily enough so that its defects could be diagnosed and cured," and who had come to be in charge of manufacturing. It has also been said that Aronson's daughter, whom Sundback later married, had something to do with the engineer leaving Westinghouse for Automatic Hook and Eye. While Sundback's electrical engineering background might appear to have been odd for someone expected to work on the development of machinery that was more in the realm of mechanical engineering, such seemingly cross-disciplinary career moves by engineers have always been common. Well into the second half of the twentieth century, there was a great deal of commonality among the different engineering curricula, with electrical engineers expected to know about machinery and mechanical engineers expected to know about electricity.

Sundback began working on improvements to the C-curity fastener, which continued to have a tendency to spring open when it was flexed, and on the machinery. After the aging Judson died in 1909, Sundback became the engineer most committed to the development of the fastener, and his new model, called the Plako because of its intended application to the seam opening in garments known as a placket. The Plako, however, also left a lot to be desired, and sales were not strong. It was said that the secretary of the company, who proudly wore a Plako in his trousers, had to rush home one evening because the fastener popped open and got stuck in that position. The Automatic Hook and Eye Company verged on bankruptcy, and it maintained its existence mainly by manufacturing various kinds of small metal devices, including paper fasteners. Sundback would not give up on the slide fastener, however, and he continued to develop the basic idea and the machinery to implement it economically with a high degree of reliability.

HOOKLESS FASTENERS

Since the hooks of the various fastener models seemed to be the cause of most of the malfunctions, Sundback began to look toward ways of eliminating them. One model, whose patent application was initially filed in 1912 and amended in 1917 (Fig. 4.4), had clasps on one side that fit over a bead on the other, with the slider opening and closing the clasps to open and close the fastener. Lewis Walker, whose financial support had been faithful for over two decades, was enthusiastic about the new model and described it as having a "hidden hook," but it came to be known as a "hookless fastener," eventually to be called Hookless No. 1. However, as could be anticipated with Howe's concept of 60 years earlier, there was considerable catching during operation and much wear and tear on the bead. Sundback went back to the drawingboard.

Sundback described the next design that he came up with as another "radical departure in principle from the design of earlier slide fasteners," one that was "built up of nested, cup-shaped members." His patent application was filed in 1914, and it represented the efforts of over 20 years of design, redesign, and development that had taken place since Judson's first promising patents were issued. The radical departure, (shown

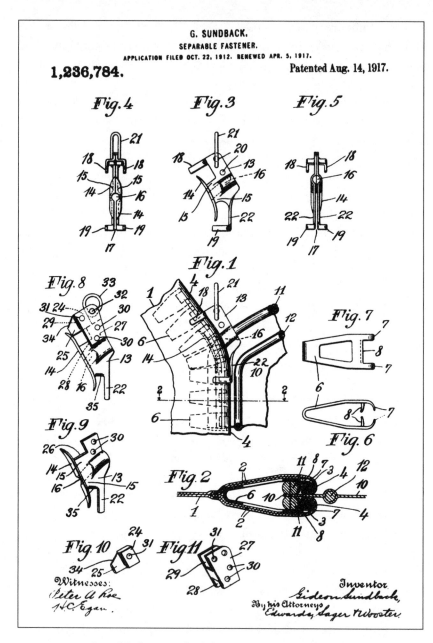

FIGURE 4.4 One of Gideon Sundback's 1917 patents for a separable fastener

in Fig. 4.5) came to be known as Hookless No. 2 and is remarkably similar to today's zipper. However, even though the principle of the slide fastener had finally been "perfected" in the latest hookless model, there remained the problem of its efficient manufacture. To address this, Sundback undertook another arduous period of development, which resulted in a new machine, which he called the S-L machine, with the letters standing for "scrapless." In its final form, the machine worked wonderfully, slicing off pieces of specially formed wire with a Y cross section, stamping a pocket into one side and letting it bulge out the other, and pinching the open part of the Y around fabric tape being fed through the machine. There was indeed no wasted or scrap metal, and production was smooth and reliable. Figure 4.6 shows a later version of a zipper-making machine.

While the long development process had finally reached its goal about a quarter century after it began, marketing and sales of hookless fasteners still faced some difficult years. There was a measure of success during World War I, when hookless fasteners were sewn into flying suits, making them windproof for flyers, and into money belts that were sold to army and navy personnel. Another application, also foreseen by Howe in 1851, was the limited use of hookless fasteners in mail pouches, but the incorporation of the devices into tobacco pouches proved to be more profitable.

Clothing applications remained scarce, in part because manufacturers would have had to invest in retraining their employees to sew in the more expensive fastening devices, and such applications were not to become very prominent until the 1930s. Rubber galoshes were another matter, however, and the hookless fastener proved to be an excellent means for opening and closing overshoes, which had to be put on and taken off very easily and quickly in cold and snowy weather. In the early 1920s the B. F. Goodrich Company began to order increasing numbers of fasteners, and they soon introduced their new product: "The Mystik Boot with the patented Hookless Fastener. Opens with a pull. Closes with a pull." The name Mystik did not attract the business Goodrich thought the boots deserved, however, and for the 1923 season they were renamed to suggest the way they zipped open and closed. Hence, the trademarked name Zippers, which soon became the unofficial name of the hookless fasteners themselves. In 1928 the Hookless Fastener Company began to use the brand name Talon to suggest the tenacious gripping power of the claw of a bird of prey and convey the idea that the newer fasteners would not fall open at the wrong time. About ten years later, the company's name was changed to Talon, Inc.

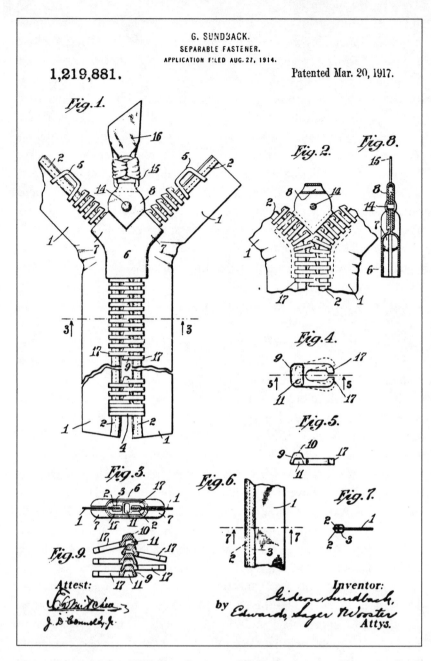

FIGURE 4.5 Another 1917 patent for a separable fastener issued to Gideon Sundback, this one resembling a modern zipper

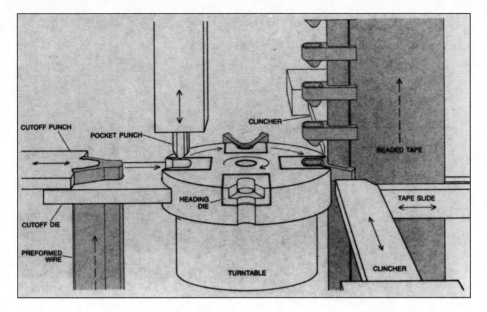

FIGURE 4.6 A later process for making zippers

RELATED DEVELOPMENTS

By the end of the 1930s, Talon was facing active competition in the zipper industry. Early patents had expired, and other manufacturers had been designing and developing their own machines. One employed by the Conmar Products Corporation stamped zipper teeth, properly called scoops, out of a flattened strip of wire at the rate of 50 per second. Another, patented in 1932 by Gustav Johnson, cast the teeth directly onto a continuous piece of zipper tape. The toothed tape was then mated with another piece, and long lengths of it were collected on spools, ready to be cut and fitted with ends and sliders and thus formed into individual zippers of appropriate size and style.

German zipper factories were destroyed during World War II, but rather than rebuild them after the war to prewar standards, the Germans developed the new technology of plastic-toothed zippers, which had been pioneered in America in the 1940s. Instead of individual metal teeth or scoops, plastic ones could be fastened to the zipper tape (see Fig. 4.7). Subsequent developments included weaving notched plastic wire into lengths of zipper and casting plastic teeth or coils directly onto the zipper tape. Plastic zippers had the advantage of being able to be made in any

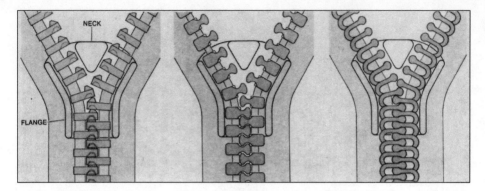

FIGURE 4.7 Forms of modern zippers, including two with plastic teeth

color. The cloth tape could be dyed to match the plastic scoops or coils, and thus the zippers sewn into garments could be made virtually invisible. This was a boon to the fashion industry and much appreciated by clothes buyers, for aesthetic and technical reasons alike.

Such developments were clearly motivated by looking for ways to make zippers better or more economical, and such incremental variations and improvements on the same basic idea characterize much of engineering research and development. In contrast to this kind of evolutionary change, however, there also can occur the kind of revolutionary change that comes not from looking at how to make an existing thing better but at how to make something in an entirely different way or based on an entirely different principle. The inspiration for such change can come to an inventor or engineer when it is least expected, but that is not to say that the individual's mind was not prepared to see in an instant the idea's potential.

In 1948, on returning from a walk with his dog through some Alpine woodland, the Swiss inventor George de Mestral stopped to remove some woodland cockleburs from his trousers and his dog's fur. As he was doing so, he wondered why the burs stuck the way they did, and he started home to look at them under a microscope in his workshop. On the way, he speculated about the mechanism that might cause the sticking, and he thought about how it might provide an alternative to the conventional zipper for fastening clothes. While he had not at the time been con- sciously thinking about inventing such a device, a few months earlier he had had an annoying time with a stuck zipper on his wife's dress. At that

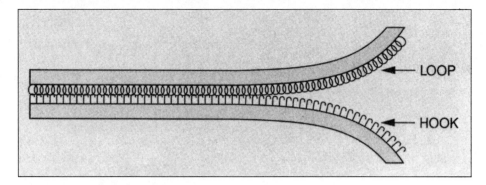

FIGURE 4.8 The principle of Velcro

time, he had wondered if he might not invent something to replace the metal zipper, but he had come up with nothing. De Mestral was no stranger to invention, however, for he had received his first patent when he was twelve years old, for a toy airplane. (He would receive his last, for a popular asparagus peeler, when he was in his sixties.)

Under his microscope, de Mestral confirmed what he had suspected, namely, that the surface of the bur consisted of numerous tiny hooks, which easily got caught in the loops of woven clothing fabric, strands of dog hair, and the like. On the other hand, when rolled in his fingers, the bur felt springy, because the fingers simply depressed the rounded backs of the hooks. Almost immediately de Mestral conceived of a new fastening system consisting of two cloth strips, one faced with tiny hooks and a mating one faced with tiny loops (see Fig. 4.8). When sewn into a dress or other garment, with the hook and loop sides facing each other, a soft but tenacious fastener that would not get stuck would result.

As with Judson's metal zipper, the basic conceptual design was sound, but it had to be developed into a smoothly functioning product that could be manufactured in a reliable way. When he approached textile experts about manufacturing the hooked tape, they were skeptical. It was only when a weaver in a textile plant in Lyon, France, produced one strip of cotton fabric with little hooks and one with little eyes, the pair of which de Mestral called "locking tape," that the idea looked realizable. Yet many details still had to be worked out if the new fastening system was to be easy to open and close, hold firmly when closed, and continue to function through many wearings and cleanings. Among the many questions that had to be faced in developing a successful device were how many hooks it should have, of what material they should be made, how they were to be formed, and so forth. Similar questions had to be addressed for the loops (A proper number of loops was eventually found to be 300 per square inch.) In time, the cotton of the working prototype produced in Lyon would be replaced with nylon, which was more durable. And among the discoveries made during the process of development was that weaving the nylon under infrared light hardened it into hooks and eyes that were virtually indestructible.

In all, it took about six years from de Mestral's conceptual design to come up with a commercially viable product and the machinery to produce it economically. The first factory to manufacture the hook-and-loop tape was opened in 1957, almost ten years after the inventor's inspiring walk. The product was sold under the catchy trademark Velcro, which was a portmanteau made by combining the beginnings of the French words "velour" and "crochet." The former, meaning velvet, refers to the soft loop tape, while the latter, meaning small hook, refers to the firmer hook tape. As with so many successful products, the name for a particular one came to be used generically. Properly speaking, Velcro-like devices are collectively called hook-and-loop fasteners, but most people continue to use the shorter and catchier term velcro. By whatever name, 60 million yards of the stuff was being produced by the late 1950s, and it soon was being used in applications as diverse as sealing the chambers of artificial hearts, holding objects in place in the weightlessness of space-craft, and closing dresses, diapers, and shoes.

Velcro was quite successful, but it did not displace the zipper in the way de Mestral might have dreamed. While the zipper continued to have its shortcomings, such as becoming stuck now and then, more significant shortcomings of velcro began to become apparent with its increasing use. No matter how hardened by infrared light, for example, the material did tend to wear out with time, especially when undergoing repeated wash-ings. Thus the application in baby diapers did not live up to its early promise. While the very noise that it makes when opened or closed is associated positively with the zipper's name, the noise that velcro makes upon being opened can be considered harsh and annoying. Another problem with velcro is its bulkiness. Whereas metal and plastic zippers had evolved toward thinner and thinner designs, so that they are hardly noticeable in clothing, velcro fasteners tend to produce a certain bulkiness, especially when applied to thin fabrics. While velcro maintains certain advantages in specific applications, it did not turn out to be the last word in fastenings.

PLASTIC ZIPPERS

Problems with metal zippers, from sticking and snagging to rusting and losing teeth, continued to attract inventors who thought they could improve upon the device. Indeed, the increasing success the zipper experienced as a commercial product during the 1930s and 1940s, with a billion a year being made by the end of that period, increasingly brought its shortcomings to the minds of inventors all over the world. One of them lived in Denmark, and his name was Borgda Madsen. He came up with the idea of a completely plastic zipper—not just one with plastic teeth or loops or scoops attached to color-coordinated fabric but one that was entirely made of plastic and that had not individual interlocking parts but a single long pair of mating grooves. Not only did Madsen's zipper remove the problems of snagging and jamming, but it had the additional advantages of being waterproof, dustproof, and airtight. As such, it had considerable potential for applications well beyond the clothing industry, but these took years of development and marketing to realize.

Inventors always have the choice of developing their own inventions and manufacturing products incorporating them, but such endeavors take money that the inventors may not have and take time that they may prefer to spend pursuing other inventions. In the case of Madsen, he sold the rights to the plastic zipper to some British investors, who in turn sold the American and Canadian rights in 1951 to some refugees from Romania. Max Ausnit, his son Steven, and his uncle Edgar formed a New York-based company called Flexigrip to exploit the new product. But first it had to be developed into a reliable product, and that responsibility fell to the youngster, Steven Ausnit, who had a degree in mechanical engineering.

Since by this time the metal zipper had become so familiar in clothing, the first inclinations of the Flexigrip developers were to promote their product as a better clothing zipper. After all, unlike hard metal zippers, it was soft and pliable and thus promised to be more comfortable. However, the plastic-grooved zipper tended to twist and come apart in such applications, and it clearly was not going to be a very successful competitor. Prior to the introduction of the plastic zipper, the conventional metal variety had also been used in such applications as garment storage bags and similar products made of vinyl. However, the conventional zippers had to be sewn with thread into these products, and the sewing holes introduced served as stress concentrators from which began tears in the vinyl that eventually grew to unacceptable lengths. Such products were not easy to repair with needle and thread, and so they were not very popular. The totally plastic zipper promised to be ideal for these applications, for it could be heat welded to the vinyl and thus provide a strong and permanent bond.

Not until the mid-1950s, however, when the Flexigrip was applied to some products that remained essentially flat in their use, was the company able to realize some measure of growth. Among its products were plastic pencil cases and plastic briefcases, and the latter became especially popular at meetings and conferences, where they were distributed to attendees to carry around the various papers and programs they accumulated. (In 1955 President Eisenhower was called upon in the White House and presented with a portfolio fitted with a plastic "toothless" zipper in conjunction with an invitation to attend an upcoming meeting of the American Society of Mechanical Engineers.) Beginning in the early 1950s, experiments with plastics other than vinyl were also overseen by Ausnit, and from the earliest days of Flexigrip there was talk of extruding fasteners out of such materials as nylon and polyethylene. The latter could be used in conjunction with polyethylene film to provide airtight and water-tight packaging that could be opened and reclosed for storage.

In the early 1960s Ausnit applied for a series of patents relating to plastic fasteners intended for the top of plastic bags, thus providing convenient storage bags for small parts and other items. His idea was to modify the way a plastic zipper would open so that it would be more effective in such applications. There was also an increasing development away from the use of a slider and toward the use of forces applied directly by the fingers to open and close the bags (Fig. 4.9), thus reducing their bulk, cost, and complexity of manufacture. Ausnit's early patents show the zipper portion of the bag to be a distinctly different assembly than the bag proper, however, and this meant that the zipper had to be heat-welded to the bag as a separate manufacturing step, with all its attendant difficulties of curling and warping of the bag material that had to be anticipated and dealt with. In particular, the bag walls and strips attached to the zipper had to be made extra-thick, and hence extra-expensive, to accommodate the heat-sealing process without forming leakage holes or otherwise dam-

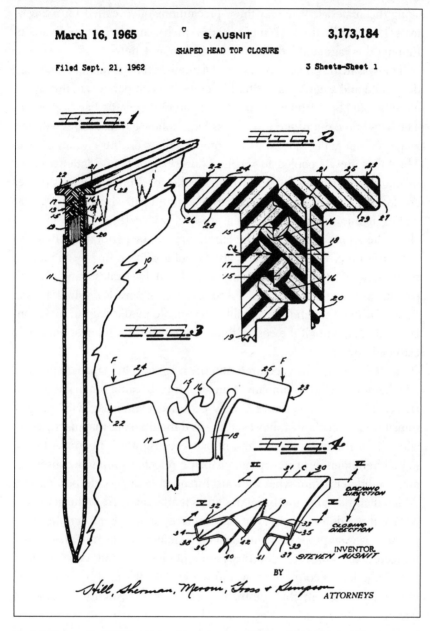

March 16, 1965   S. AUSNIT   3,173,184

SHAPED HEAD TOP CLOSURE

Filed Sept. 21, 1962   3 Sheets—Sheet 1

INVENTOR.
STEVEN AUSNIT
BY
Hill, Sherman, Meroni, Gross & Simpson
ATTORNEYS

FIGURE 4.9 One of Steven Ausnit's many patents for a plastic zipper, this one opened and closed either with a mechanical slider or manually

aging the material. Whereas these precautions often called for a bag at least 3 to 4 mils thick (1 mil = 0.001 inch), an alternative means of forming the bags enabled them to be as thin as 1 mil.

This was made possible because a Japanese inventor, Kakuji Naito, had developed and patented a method whereby the components of the zipper closure could be extruded as an integral part of the plastic bag (Fig. 4.10). The bags come out of the extruder as long hollow circular tubes with the zipper components located at the proper points on the circumference. The tubes are air cooled to set the plastic before being flattened, thus mating the zipper parts, and rolled onto collector drums. The bags proper are formed by unrolling the flat tube, printing it where desired, and cutting it into bag lengths, which are heat sealed (Fig. 4.11). The top above the zipper opening can be left uncut, can be perforated, or can be cut at this time. The bottoms can also be cut to allow filling by automated machinery, after which they can be reclosed by heat sealing. Naito's patents were assigned to the Tokyo-based company Kabushiki Kaisha Seisan Nihon Sha, and they enabled resealable plastic bags to be manufactured at about half the cost of those made by heat-welding a separately extruded zipper.

In 1962 Ausnit's firm acquired American rights to the Japanese process, and the newly named company, Minigrip, Inc., became the first to manufacture in the United States a fully extruded plastic bag with integral miniature zipper. At first, however, it was difficult to get manufacturers to adopt it for use in their products, in part because it was unconventional. (The phenomenon of new products being rejected simply because they are too different from what they are intended to supersede has led to a design dictim among industrial designers that is captured in the acronym, MAYA. It stands for "Most Advanced Yet Acceptable.") For example, when the new bag was proposed as an ideal reclosable, dust-free outer packaging for record albums, record industry representatives rejected it because they did not believe record buyers would understand the package and would cut or tear it open, thus destroying its relatively expensive reclosable feature.

The way around this impasse came when Minigrip, in addition to making and selling resealable plastic bags themselves, granted the Dow Chemical Company an exclusive license to sell them directly to consumers through supermarkets. These handy new products came to be known as Ziploc bags, and their success helped Minigrip market the more heavy-

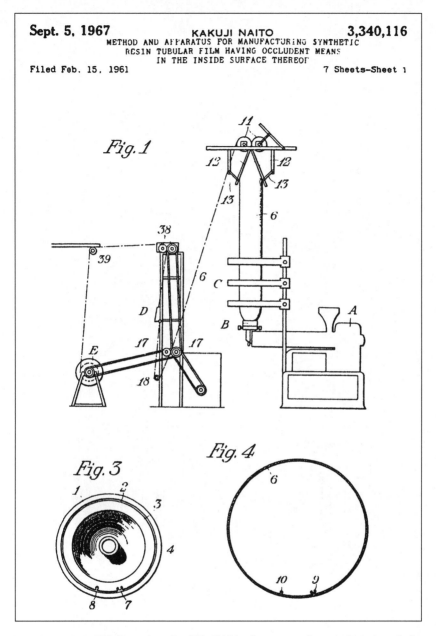

Fig. 1

Fig. 3

Fig. 4

FIGURE 4.10 U.S. Patent issued to Kakuji Naito for a means for manufacturing tubular film having an integral plastic zipper

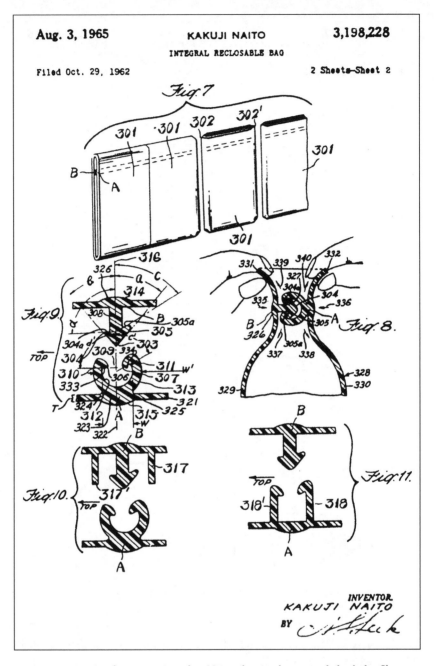

FIGURE 4.11   Another patent issued to Naito, showing how extruded tubular film may be slit for filling and cut into individual bags

One of the frustrations experienced by users of early reclosable plastic bags was that it was not easy to determine when they were closed securely. A competitor thus introduced the idea of making the two sides of the plastic zipper closure appear as stripes of different primary colors, one yellow and one blue, which, when properly mated to give a good seal, produced a uniform green band. This useful improvement was not only patentable but also provided a very effective marketing device.

Can you think of some other modifications of plastic bags that would make it easier to close them tightly and to know that they are closed?

duty bags to commercial and industrial users, a right the company had retained.

The success of Ziploc bags naturally attracted rival brands, which employed improvements on the basic design to secure separate patents. As with the evolution of all artifacts, arguments for these new patents rested upon finding fault with existing patents. Ironically after reclosable bags became commonplace in the kitchen and workshop, it was not opening them but closing them properly that became the focus of manufacturers and users alike.

But not all potential competitors looked for new patents as a means of entering the market. Manufacturers in Taiwan and other Far East countries, in particular, totally disregarded the patents that Ausnit and Minigrip had so systematically acquired in order to protect their investment. Plastic bags from Taiwan, for example, produced with inexpensive labor and not having to recover the research and development or patent licensing costs normally associated with a new product, could be sold for a fraction of the cost of the Minigrip product. In such cases of unfair trade, a company can appeal to the International Trade Council Court, which Minigrip did. Such appeals are seldom upheld, but in this case an Exclusion Order was issued by the Court which essentially banned bags of foreign competitors that infringed on the patents held by Minigrip.

The stories of the original zipper, Velcro, the plastic zipper, and the resealable plastic bag derived from it each span many years and show how long and arduous the development of a conceptual design or a patent idea can be. These case studies also demonstrate how the success of one product leads to the conception and development of many derivative ideas, which in turn lead to others.

5

ALUMINUM CANS AND FAILURE

An idea that unifies all of engineering is the concept of failure. From the simplest paper clips to the finest pencil leads to the smoothest operating zippers, inventions are successful only to the extent that their creators properly anticipate how a device can fail to perform as intended. Virtually every calculation that an engineer performs in the development of computers and airplanes, or telescopes and fax machines, is a failure calculation. In analyzing the cantilever beam, even Galileo began by making assumptions about how it would break or fail. Today, in designing a cantilever bridge, the engineer must have an understanding of how much load the individual steel members can safely carry before they pull apart or buckle and how much deflection can be allowed in the center of the bridge.

Such considerations, made explicit at the outset, are known as failure criteria, and they provide limits that cannot be exceeded as the design develops, whether the artifact is a bridge, a building, or a beverage can. As the engineer calculates the forces and deflections of a trial design, each resulting numerical calculation takes on meaning and becomes acceptable only in comparison to failure criteria, which may have been determined by careful laboratory experiments on the materials and components in question. While most of our discussion up to this point has been couched in terms of the analysis of force and strength, form and function, similar remarks apply to problems involving other properties and criteria such as heat transfer calculations and the melting point of materials, or voltage and current calculations and their safe values in electrical conductors.

What distinguishes the engineer from the technician is largely the ability to formulate and carry out the detailed calculations of forces and deflections, concentrations and flows, voltages and currents, that are required to test a proposed design on paper with regard to failure criteria.

The ability to calculate is the ability to predict the performance of a design before it is built and tested. By understanding how and why a proposed design can fail, and by being able to calculate the quantities needed to assess whether failure conditions prevail, the engineer is able to test a design on the drawingboard or on the computer screen before any steel is erected, any valve is opened, or any switch is thrown. Calculations that indicate failure conditions in the design enable the engineer to modify and remodify the design until it is ready to be realized.

Failure manifests itself differently in different branches of engineering. In environmental engineering, for example, where limiting conditions may be expressed in parts of pollutant per billion parts of water, an environmental engineer needs to know how to calculate the amount of a contaminant that might seep from a proposed hazardous-waste disposal site into the soil near a source of groundwater, and then compare this result to the criteria set by federal regulations. Some problems of engineering design do not lend themselves so much to analytical calculations as to trial-and-error or to build-and-measure techniques.

Sometimes engineers test their ideas by simply imagining scenarios of use and behavior that might lead to the failure of a product and then trying out the product under those conditions. In the design of computer programs, for example, the software is first "alpha tested" by its designers and then "beta tested" by real users to uncover the bugs that might have been inadvertently introduced during design or modification, and to discover how the program might fail to perform as intended. Whatever the method used to test a design, obviating failure is always the underlying principle.

Failure can take nontechnical forms. A design can be considered a failure if it is environmentally unsound or aesthetically unsatisfying. Such criteria ought to be taken into account from the beginning of the design phase, just as the strength of materials is. Often, however, technical details dominate the early stages of a design, when an engineer is trying to establish whether a particular idea is physically feasible.

In order to explore in more detail the role of failure in successful engineering design, let us consider something a bit more complicated than a paper clip or a pencil but still less massive than a bridge or a water supply system, and more tangible than the flow of electrons through a wire or a computer chip. All of us are likely to have held an aluminum beverage can in our hand at one time or another, and so we all may be assumed to have at least a passing familiarity with this now- ubiquitous artifact. The aluminum can may be described as a pressure vessel when it contains a carbonated beverage (and even more so when it has been dropped or shaken up), and in this regard the lowly can has to be designed as carefully against accidental failure by explosion as does a steam boiler or a scuba diving tank. The successful design of an aluminum can depends on understanding how it can fail to contain its contents and on obviating the possibility that it will fail before it is supposed to. Carbonated beverage cans are made safe and reliable by using enough material in the proper configuration to keep the intensity of force within bounds of strength and to keep the cans from bulging out in the wrong places. But engineering also has to a lot to do with economics, and the object in designing and manufacturing something made by the billions, like beverage cans, is to make them extremely safe and reliable while at the same time costing as little as possible. A too-expensive beverage can will fail to survive the competition. While this may be considered a nontechnical failure mode, it is a failure mode nonetheless.

THE ALUMINUM BEVERAGE CAN

The earliest food and beverage cans were made of iron and were often as heavy as the food they contained. Moreover, their robust construction made opening them a major effort—some of the first iron cans even had instructions for opening that involved the use of a hammer and chisel. With the development of stronger steel, cans could be made thinner and thus lighter, but still they were difficult to open, and therefore specialized can openers were invented to make the task easier. Aluminum—because it was more expensive than steel—could not compete as a material for food cans, which had to be strong enough to resist being dented too easily. But with beverage containers it was a different story. Because soft drinks and beer pressurize a can, they provide some of its stiffness and make a thinner (and cheaper) can wall possible. On the other hand, the contents under pressure in turn requires the can wall to be strong enough to resist being split open just by the force of its contents.

Since aluminum is generally a much more ductile material than steel, it can be formed into containers in more direct and effective ways. Steel cans were long made by bending a flat sheet into a hollow cylinder, joining it along its seam, and adding top and bottom, a multistep, multipart process. By contrast, the entire seamless bottom and sides of an aluminum can could be formed from a single disk of metal (see Fig. 5.1), with only a top to be added after the can was filled with its contents. This formability of aluminum gave it a clear advantage in making cans by the billions.

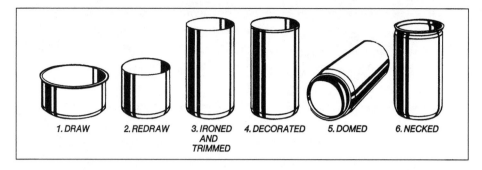

FIGURE 5.1 Steps in forming an aluminum can body

It is relatively easy to calculate how thick the aluminum must be to contain the pressure of the beverage and to take advantage of the pressure as it works to stiffen the thin-walled can against being crushed or dented. But the pressure would also tend to balloon out the flat bottom of an aluminum can, preventing it from sitting flat on a shelf or table. Thus, the characteristic inwardly dished bottom was developed, to act somewhat like an arch dam against the pressure while allowing the can to sit flat and stable on its rim. Were it not for this and other of its various structural features, the aluminum can would fail to be a very useful innovation.

The first aluminum cans were opened in conventional ways, which usually meant with a pointed can opener known as a church key. When this once-indispensable implement was first introduced, steel beer cans came with illustrated directions for its use printed on their labels. By the time aluminum cans were introduced, there was no need for such directions, and using the common opener was not considered especially inconvenient, since people had long become accustomed to the two-step process: First, a triangular-shaped hole was punched with the lever-opener in the can top, and then the whole can was rotated 180° about its cylindrical axis so that a second hole could be punched diametrically opposite the first. The contents of the can could then be poured or drunk easily from one hole, while the other allowed air to enter the can and displace the liquid, thus providing a more steady flow without spilling any of the contents. The two-step process soon had become such a familiar one to can users that they hardly gave it a second thought, and few probably realized it could be improved upon.

However, the necessity of using a church key did present a problem to at least one thirsty individual when a can opener was nowhere to be found, and little annoyances and predicaments such as this are what catch the attention of inventors and clever engineers. One evening, while Ermal Fraze, of Dayton, Ohio, was reflecting on having found himself earlier that day without a can opener while on a picnic, he set out to devise a self-opening can. Though many may have realized how convenient it would be to do away with can openers, Fraze was in a position to do something definitive about the problem because of his knowledge of metal forming and scoring. He came up with the idea for the now-familiar aluminum tab-top or pop-top can in an evening and perfected it shortly thereafter (see Fig. 5.2).

Although much of engineering has to do with avoiding failure at almost any cost, the development of a self-opening aluminum can top presents an interesting example of balancing the competing aims of preventing and encouraging failure. Clearly, we would not want a beverage can to pop open spontaneously or too easily, and so the pop-top must be a relatively robust design. On the other hand, we do not want to have to exert undue force to open the can when we are thirsty. Scoring the can top just the right amount leaves enough strength to contain the pressure while at the same time providing a preferred site for the metal to fail (that is, tear open) when desired. The tab attached to the top of a pop-top can effectively serves as a small lever (somewhat akin to a little church key) to magnify the force of the fingers and cause the top to fail in a controlled way. However, because the contents are under pressure, the process is complicated.

In Fraze's early designs, the tab was riveted to the can top, and the rivet served as the lever's fulcrum. In some early designs, the rivet was a source of leaks. In others, the rivet was too easily pulled out of the top, thus leaving the thirsty consumer scurrying to find a church key to open the can in the conventional way. In other cases, when the lever was pushed into the can top to break the pressure seal, the rush of pressure was liable to make the top fly off and threaten the well-being of the would-be drinker. Such failures of the device to work properly led inventors and engineers like Fraze to improve incrementally the pop-top until it was so reliable and operated with so few surprises that it began to be taken for granted.

One engineering professor uses the aluminum beverage can in a very dramatic experiment to demonstrate the complexity of failure modes that engineers often face in their pursuit of fool-proof designs. He places an unopened can inside a large plastic Ziploc bag and, to be doubly sure, for reasons that will be apparent, encloses the can and bag in a second Ziploc bag. He then places the can between heavy rings on the platform

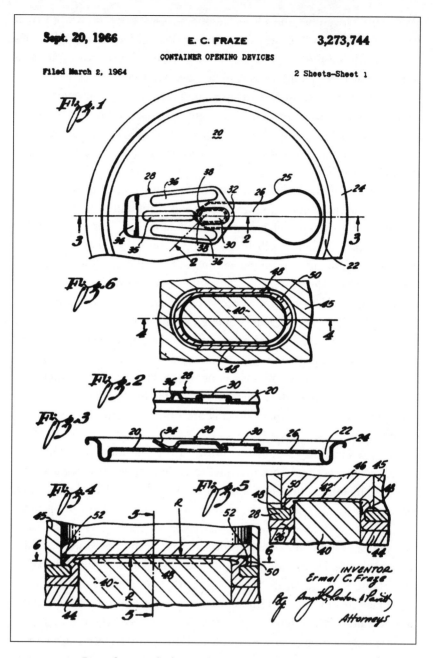

FIGURE 5.2 Patent for an early design of an easy opening can with removable tab top

and moveable head of the kind of testing machine usually employed to crush concrete samples to determine their (failure) strength, and the kind that David Letterman used on his old *Late Night* show to crush everything from light bulbs to watermelons.

In the aluminum-can engineering experiment, the rings bear on the rims of the can so that the machine's force does not bear directly on the center of either the top or bottom of the can, which can be taken out of the machine at various stages during the test to be inspected for damage. Before turning the machine on to begin the slow but deliberate crushing of the can between the powerful faces of the machine, the students are asked how they think the can is going to behave. In other words, how is the can going to fail? The variety of answers that result demonstrates how much more complicated a structure the aluminum can is than a cantilever beam. Whereas Galileo was faced with the relatively easy but, in his time, far from trivial problem of figuring out how the cantilever failed at the wall, there was no doubt about where it would fail.

In the case of the aluminum can in the testing machine, however, there are many conceivable ways in which failure can occur, and students usually have little trouble coming up with a goodly list. Furthermore, different failure modes can develop as the test progresses. Among the more obvious possibilities are: (1) the pressure of the compressed liquid splits the can's sides open, (2) the compression of the top causes the stepped neck to be pushed down into the can, (3) the can's sides wrinkle the way they do in an empty can, (4) the can's bottom pops out to accommodate the increased pressure, (5) the can's bottom splits open, (6) the can's top arches to accommodate the pressure, (7) the rivet in the pop-top is ejected or split open, acting like a pressure relief valve, (8) the top cracks open where it is scored, (9) the can begins to leak around the rim where the top joins the sides. Different styles of can are likely to exhibit different failure modes in different sequences and at different load levels, and the plastic bags themselves may fail in their own way, giving the onlookers a sticky shower. (To avoid such surprises, it is best also to surround the can and bags with a heavy plexiglass cylindrical shell.)

Whatever happens to a can in a testing machine, it is clear that the possibilities of failure are multifarious and generally beyond deterministic calculation. That is not to say that aluminum beverage can manufacturers and others do not consider such things. Among the losers as aluminum cans have become widely accepted has been the steel beverage can (see Fig. 5.3), and the steel can industry has long been trying to regain its

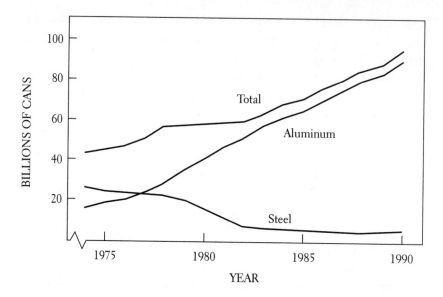

FIGURE 5.3 Growth of aluminum beverage cans at the expense of steel cans

once-dominant position. For years, research and development has been conducted to come up with a competitive steel can, and among the most vexing of problems has been how to design a steel pop-top that does not leave a sharp or jagged edge to cut the drinker's lips. For a while, steel can bodies were fitted with aluminum pop-tops, but this complicated recycling efforts.

By 1993 steel cans were being made, like aluminum ones, in a drawing process that produced walls as thin as 0.0025 inches, which is less than the thickness of a piece of copier paper. Even with such thin walls, the "column load strength," or the force it would take to crush a full can by standing on it, was maintained at 340 pounds, and the steel can's "dome reversal strength," or the pressure it would take to pop out the bottom, was held at 100 pounds per square inch (psi). Even with such advances, the approximately 3 billion steel beverage cans shipped annually amounted to only a small percentage of the 100 billion total beverage cans, about 97 percent of which were aluminum, made in the United States.

Calculations employed in the design of beverage cans, or of engineering systems even more complicated, provide quantitative starting points for assessing if failure criteria are met, and guidance for changes in response to unacceptable behavior. For example, knowing the highest

Some years ago, in a popular commercial on television, a fellow smashed a beer can against his forehead. Did it hurt? Was the can closed or open, and would that make a difference? (This question requires only a theoretical answer. Do not attempt to investigate through experimental means.)

Although aluminum beverage cans must be designed primarily to withstand the internal pressure of their contents, most are designed to be strong enough (before being opened) to support a good-sized person standing on them. Is this behavior something that happens frequently in fraternity houses and such places, or are there other reasons for this design criterion?

pressure the contents are likely ever to be under, one could calculate the thinnest aluminum wall that could contain such pressure. To make a can thinner would not be a very likely place to start. But the thinnest possible can that could contain the pressure might not be acceptable, for it might not support enough vertical load to allow cases of cans to be stacked for shipment or storage. Further, there are other failure modes (such as those in the above list) and other considerations that relate to the use of the can when it has been opened and the pressure has been released.

Because an opened (and thus unpressurized) aluminum can must be sufficiently rigid to stand up on a table and not be crushed in the hand that lifts it to the lips, there is a limit to how thin it can be. We all know that cans are already almost at their practical limits of thinness because we must be careful not to hold them too tightly lest their sides buckle in and the can becomes unstable in our grasp. While such a consideration might not be strictly a question of strength or safety, it is clearly a limiting condition for the can to function comfortably in use, and it certainly constitutes another failure criterion that must be taken into account in the design.

An analogy can be found in building design. Much taller and more slender skyscrapers than presently exist could easily be built economically without any danger of collapsing, but with height and slenderness come flexibility, and this becomes the limiting factor on what can be practically achieved. The top floors of too flexible a skyscraper could sway several feet in just a gentle wind, and this can cause coffee to slosh around in mugs, elevators to bang and bind in shafts, and office workers to feel queasy at their desks. While no dramatic or catastrophic failure is occurring in such a situation, the structure has clearly failed to provide psychologically or physiologically effective office space for its users. From the total engineering perspective, and the economic or use value of the structure, this could be as much a failure as if the building had to be abandoned for structural reasons.

ENVIRONMENTAL FAILURE

By the early 1970s it became evident that the removable tab-tops of beverage cans were creating an environmental crisis. The small, sharp, ringed tabs of aluminum were being disposed of by the billions on roadsides, in parks, and on beaches. Besides creating a litter problem, they were presenting physical dangers where recreation seekers went barefoot and where small children swallowed things not intended to be ingested. Especially on beaches, where the tab-tops were often too small to be caught in the standard beachcomber's rake, there was many a foot cut on an unseen tab lying just below the surface of the sand. Conscientious drinkers began to drop the tab into the can from which they were drinking, only to swallow it with a mouthful of beverage. In short, what began as a technological godsend for drinkers without church keys ended up as a devil of a problem.

This was a clear failure of the pop-top can to be a benign beverage container, and it sent inventors and engineers back to the drawingboard. A plethora of solutions resulted, including a can developed by the Adolph Coors Company in which a small aluminum button was first pushed into the top to break the pressure seal and then a second, larger button was pushed in to provide a drinking hole. Both buttons remained hinged to the can top and so created neither a litter problem nor a hazard. However, in order for the can top to function properly (actually, to fail in the proper mode at the proper time), the little button had to be small enough so that the total force of the pressure acting on it did not require too great a push to break the seal. But the smaller the button, the sharper it felt to the finger pushing it, and so opening a can could be an uncomfortable process. Such annoying details, coupled with the fact that two separate buttons had to be pressed to get at the beverage, as if one were using a church key again, led other inventors to look for alternatives. What evolved is the now familiar pop-top, in which the manipulation of a lever breaks the pressure seal, opens an orifice, and folds back (but does not break off) a flap of aluminum.

The stay-on tab was the invention of Daniel F. Cudzik, of Richmond, Virginia, who worked for the Reynolds Metals Company. In 1976 he was awarded a United States Patent (Fig. 5.4) for his invention of an easy-open aluminum can top with a tab that stayed attached. This development was instrumental in keeping the aluminum can from being outlawed altogether on environmental grounds. The new top was variously called an "ecology" top and an "environmental can end," the former term being especially popular in the 1970s. Reynolds Aluminum Company promoted its new stay-on-tab can end the way most new technologies are introduced—by comparing it with what it improved upon. However, also like most novel technologies, the way the new can top worked was not necessarily obvious to the potential customer, and so early promotional cans

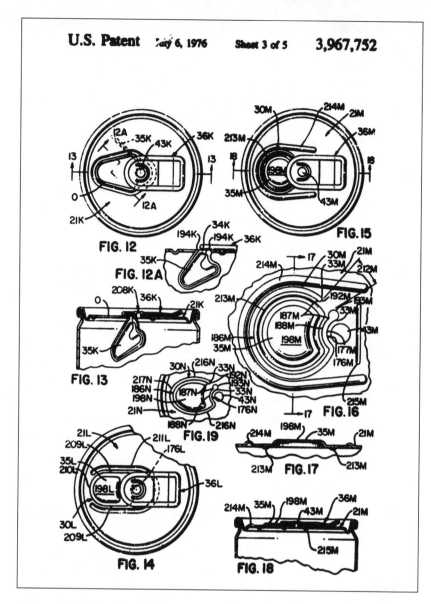

FIGURE 5.4 Patent drawings for a can top with a stay-on-tab

were imprinted with instructions for opening in much the same way the first cans intended to be opened with a church key had carried printed instructions.

Before the idea of the new product could be effectively sold to beverage companies, however, they wanted to know how the consumer might receive it. Thus, Reynolds Aluminum conducted consumer studies of the new can design in ten supermarkets in Fort Pierce, Florida, and a control study in a similar market in Cocoa, Florida, where only conventional cans were available. By asking purchasers of the new-style cans to rate them in various categories, Reynolds concluded that the stay-on-tab had "overcome the major disadvantage of . . . the throw away tab. Stay-on-Tab is perceived to be better for the ecology." This and other positive comparative results led to company to be optimistic about "the design's viability and potential for future success."

In the sixteen years following the issuance of Cudzik's patent, which included much legal maneuvering regarding patent infringement, almost a trillion such tabs have been manufactured by Reynolds and other companies licensed to use the idea. The stay-on-tabs alone amounted to over four million tons of aluminum that was recovered and recycled rather than discarded on roads and beaches to pose litter and safety hazards. The now-familiar tab opening on beverage cans, like all artifacts, has its shortcomings and fails to function as fault-free as one might imagine, but we have adapted to it, as we do to most artifacts, and use it almost unthinkingly. Because it is not perfect, there are likely to more than a few Frazes and Cudziks working on new inventions to remove the shortcomings of the old.

No matter how the cans are opened, the waste produced by packaging each twelve-ounce portion of soda or beer in its own aluminum pressure vessel appears to be a terribly wasteful use of energy and other nonrenewable resources and a litter-producing practice to boot. From the introduction of the aluminum can, manufacturers recognized that in order to make their product competitive with their steel counterparts, they would need a steady and reliable source of the metal. Recycled cans are a most attractive and dependable source of supply. It is said that the recycling system is now so efficient that the aluminum in a used can may show up in a new one in as little as six weeks. More than six out of every ten aluminum cans are now being recycled, and so the aluminum used in the beverage industry may be considered, if not a renewable, at least a recoverable resource.

The cost of aluminum has been a major factor in driving engineers to find ways of making beverage cans as light as possible. This quest, known in the industry as "lightweighting," resulted in the average can weight being cut almost in half in the period from the mid-1970s to the early 1990s. One obvious way to reduce a can's weight is, of course, to make its wall thinner. But there is a clear limit to how thin a can can be, to be both strong enough when closed to contain the pressure and stiff enough when opened to be able to be held comfortably in the hand. Thus, alternative ways of lightweighting the can were sought.

Among the most effective means of reducing weight was in reducing the size of the can top. Because a top must be thicker than a can's body to maintain strength after being scored and riveted, the top takes a disproportionate amount of aluminum to fabricate. Thus, in the mid-1970s, can manufacturers began to narrow the can body ever so slightly in order to employ a smaller-diameter top. Since the area of a can top is proportional to the diameter squared, a small decrease in diameter resulted in a substantial reduction in aluminum used. However, the can body cannot be narrowed too much, or else it would not feel right in the drinker's hand. To keep the can body sized for comfortable use and to maintain familiar proportions, the body began to be tapered at the top so that a smaller top could be fitted to it. By the late 1980s, this tapering had become very pronounced, but it could only be taken so far without making the can look too unconventional or be too difficult to drink from. Furthermore, there was the need to have the top diameter large enough to incorporate a tab opening.

With can lightweighting by reducing top diameter reaching a practical limit, can manufacturers began to look for other ways to reduce the amount of aluminum needed to make a can, and they returned to the problem of reducing wall thickness. A new wall design, fluted like a classical column, was at first considered for aesthetic reasons. When it was found also to increase the strength of the can by 20 percent, it was looked to as a new means of reducing can weight by as much as 10 percent. The fluted can did not seem to fare well in the marketplace, however, and its appearance in the mid-1990s was short lived.

In 1996 Coca-Cola announced a new can shape that seems to sacrifice lightweighting for the sake of appearance. The new contoured aluminum can with a wavy body and a larger diameter top was designed to evoke the shape of the classic green-glass Coca-Cola bottle. The can's faceted sides gave it strength against the pressure within, and the contour of its body enabled the can to be held with a lighter grip by a drinker. A similarly contoured can, but larger and made of a steel body with an aluminum pop-top, was introduced years earlier in Japan to contain Sapporo beer, which Japanese consumers were accustomed to drinking from bottles. In spite of its untraditional shape, the Sapporo can was not well received in Japan, and it is now used only for export. Whether the new Coca-Cola aluminum can succeeds will depend similarly upon cultural factors.

Engineering has many dimensions, but the idea of failure spreads across all of them. Understanding how a can may fail to work structurally as a pressure vessel is only one aspect of the engineering problem associated with manufacturing and distributing beverages in light, inexpensive, convenient containers. By appreciating how these containers may also fail to function aesthetically, environmentally, ergonomically, or in any other way, we can better understand the multifarious arenas in which engineering problems must be conceived and attacked.

6

The technical development of a reliable zipper took pure and persistent engineering, supported by the financial backing of loyal and patient investors. Many inventions have survived equally difficult and uncertain gestation periods, complicated in large part by the fact that they could not just be introduced into a marketplace that was not prepared for them. Thus, the light bulb was of no use without a reliable and effective supply of electricity upon which its operation so clearly depended. Indeed, it was Thomas Edison's recognition that a light bulb filament had to have a high electrical resistance in order to keep down the cost of the power supply network into which it would be plugged that led him to look for a proper filament material. He tested and discarded thousands of potential materials before coming up with the right combination of electrical resistance and incandescence. When asked why he was not discouraged by the string of failures, Edison is said to have responded that, rather than discouraging him, each failure spurred him on with the new information that another material did not work and so could be disregarded.

The automobile, no matter how well engineered and technically advanced, would not be a very effective vehicle were it not for the network of roads, including the bridges and tunnels carrying them over water and through mountains, that developed along with the horseless carriage. Imagine how different a machine the automobile would have to be if there were not a string of gasoline stations across the country. And imagine televisions without broadcast or cable networks. Such contexts, which develop along with inventions and in which inventions develop into innovations, involve systems and infrastructure which shape technology and its use. Indeed, after a while the individual artifacts and the supporting infrastructures become so interdependent that the one no longer can be considered without the other. Thus, airplanes are pretty much useless without airports, and who would have thought of building airports if there were no airplanes. Of course, what exactly constitutes an airport at any stage in the airplane's development could be a matter of definition. The sandy beaches at Kitty Hawk effectively constituted an airport for the Wright Brothers.

What comes first in the development of technological systems can clearly be a chicken and egg question. In the case of the airplane, once there is a costly network of airports and a complex system of reserving seats on scheduled airline flights between them, technical changes in the airplane itself cannot be made without considering the entire system in which it operates. It would be as foolish, for example, if aeronautical engineers proceeded to develop new energy-efficient jumbo jets that required runways three times as long as existing ones or airport terminals with much greater capacity than they currently have. It would be equally foolish for new airports to be designed with runways shorter or waiting areas smaller than those required by existing aircraft.

Technology is the catch-all term used to describe objects and the networks, systems, and infrastructures in which they are embedded, as well as the patterns of use that we impose upon them and they upon us. Technology is clearly context-dependent and ever evolving, and engineers play a central role in influencing how it develops. Engineers design the basic parts into functioning artifacts, and they also necessarily become involved in designing the supporting networks and systems. Merely having an idea, patenting it, and securing financial backing is almost never enough to ensure that an invention will become part of current technology; invariably a further engineering effort is required to bring a concept to usable reality, as case study after case study reveals. Thus we might say that engineers not only affect but effect technological development.

This is not to say that engineers control technology or are totally responsible for its being what it has been, is now, or will become, but they clearly are in the middle of the action from start to finish. Every artifact and technical breakthrough has a history that illuminates the general nature of this engineering endeavor, and we can fully understand why a thing looks and works the way it does only by understanding its development in the context of its times, technology, and culture.

THE FACSIMILE TRANSMITTAL MACHINE

The now-ubiquitous fax machine presents an interesting case study of the importance of context. Clearly, sending a fax is a meaningful thing to do only if there is a compatible fax machine at the receiving end. The idea for transmitting written, printed, or graphical material over long distances is not new, as demonstrated by such ancient notions as winged messengers and carrier pigeons and the more modern concept of the postal system. With the discovery of electrical phenomena and the invention of the telegraph and, later, the telephone, the transmission of words and word pictures through other than an animal or mechanical medium became a reality. Indeed, in the second half of the nineteenth century telegraphic messages could be carried over transoceanic cables between hemispheres. Telegraph offices served as terminals to and from which young messengers like Pittsburgh's Andrew Carnegie carried printed documents. A message was encoded, transmitted, decoded, and put into paper form again for delivery by other messengers to their ultimate destination. The clear advantages of transmitting words in such a way was obvious to everyone, and especially to inventors. It was also obvious that one of the system's greatest shortcomings was that pictures and graphical materials could not be transmitted directly. Such limitations of a technology sow the seeds of invention.

The story of the development of the modern fax machine is a long and instructive one. In 1843 a British patent for a method to transmit images electrically was granted to a Scots clockmaker named Alexander Bain. Bain was a prolific inventor who perfected the electric clock in 1840 and devised the inked typewriter ribbon in 1841. His inventive interest in both the electric telegraph and the synchronized electric clock prepared him to see how to combine their features into a pair of devices that could send and receive graphical material. As with the early versions of many inventions, Bain's facsimile machine was crude by today's standards. The graphical material to be transmitted had to be prepared as a raised image on a metal block, much the way the reverse images of letters are raised from typewriter keys. A stylus was passed over the block and rose and fell in conformity with the raised image, thus breaking and completing the transmission circuit (Fig. 6.1). A synchronized stylus on the receiving machine was then raised and lowered to reproduce the image. By incrementally repositioning the transmission block in a second direction, the entire image could in time be reproduced. Bain's basic idea of scanning an image is preserved in modern fax machines, but the clear disadvantages of his first efforts drove subsequent inventors to seek ways of simplifying his cumbersome machines and reducing the effort needed to prepare images for transmission.

Among the first improvements was to replace the bulky, costly, and

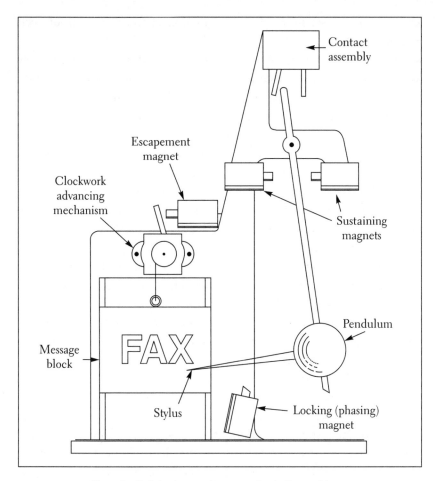

FIGURE 6.1 Alexander Bain's nineteenth-century facsimile machine

difficult-to-prepare metal plates with tinfoil on which the image to be transmitted was drawn in ink. As the stylus passed systematically over the tinfoil, the current through the circuit was altered, and so the information could be transmitted electronically, to be recreated a line at a time on treated paper by a stylus incorporated into the receiving circuit. The first commercial facsimile system employing such a technique comprised a pair of machines installed in Paris and Lyons in 1865 by Abbé Caselli. No less an inventor than Thomas Edison was interested in improving the new technology, and he attacked the important problem of keeping the sending and receiving styluses in phase and synchronized by employing elaborate gearing and electromagnetic pendulums at each machine, which could be set in synchronized motion at the beginning of each transmission.

One of the limitations of using styluses was that they were principally on-off devices. This was fine for transmitting line drawings and graphs, but it was not effective in communicating the gradations of black and white that were contained in photographs, and so their images could not effectively be sent electronically. Near the end of the nineteenth century, however, the photoelectric cell was developed, and this enabled photographic images to be scanned and converted into a spectrum of electronic signals, thus transmitting shades of gray. As an added advantage, images could be sent faster, and Arthur Korn, a German inventor, first demonstrated the technique successfully in 1902. The transmission of photographs electronically had an obvious commercial appeal to newspapers, who were thus willing to invest in the new technology, and a wirephoto circuit was established by 1911 to connect offices in Berlin, London, and Paris. Networks were established in Europe and the United States after World War I, and transatlantic transmission of photographs was established in the 1920s. In the early 1930s, the *New York Times* supported the work of the inventor Austin Cooley, who was developing a fax machine that was small enough to be carried by an individual to transmit news pictures via ordinary telephone lines.

One of the weak links in the chain of any newspaper publishing enterprise is the printing and distribution network. The latter is especially capital and labor intensive because of the fleet of trucks and army of workers it requires to get the news delivered before it is stale. Newspapers have always been vulnerable to being shut down by strikes of printers and truckers, and a technological means of circumventing that had a clear appeal to publishers. The promise of being able to transmit facsimiles of newspapers around the clock to radio receivers in homes and offices led to the establishment by 1940 of 40 stations sending out experimental newspapers via ultra-high-frequency transmission. These experiments were interrupted by World War II, and the emergence of television in the wake of the war led to the abandonment of commercial fax broadcasting by newspapers.

There were obvious wartime advantages to having fax machines for sending such documents as terrain and weather maps, as well as photographs, to front lines, and so the military continued to be a viable market for the technology. Among the few commercial efforts that succeeded in the 1950s was that of Western Union, which sold Desk-Fax units that could send and receive telegrams without requiring anyone to leave the office. This technology was attractive only to those customers whose business volume justified investing in the equipment, however.

The very first telegrams had to be sent by Morse Code and decoded by a human ear and recorded by a human hand at the receiving end. The teletypewriter was developed to automate the typing of incoming messages, and early versions produced a continuous line of text on a paper tape. This was cut into lengths that could be pasted on the familiar yellow Western Union telegram forms that can be seen in old movies. Teletypewriters evolved into instruments that could be used also to send messages, and by incorporating such features as line-end functions into the keyboard, messages were received in a form more akin to a typed page. Telex, short for Teletypewriter Exchange Service (TWX in another version), was a system that employed telephone lines as multichannel conduits to transmit data between teletypewriters. This now largely superseded technology retains a vestigial presence on some business letterheads that continue to list Western Union Telex or TWX addresses, especially those which were aptly personalized, such as the Telex address CIVILS for the British Institution of Civil Engineers. The Telex and TWX services were combined in 1970, but not long thereafter fax numbers began to be added to letterheads, in most cases replacing the numbers and letters of the older technology.

Facsimile transmission, albeit relatively crude and slow, was thus well established in specialized uses long before fax machines became the familiar instruments they are today. It is instructive to understand why a century of engineering and technological advances were not alone sufficient to give the system a more widespread presence. After all, typewriters and telephones were also crude by today's standards, and yet they pervaded places of business of all sizes and in all locations.

TELEPHONE NETWORKS

Among the reasons fax machines were not more widely incorporated into offices earlier than they might have been, at least in America, has to do with the available communications infrastructure that had developed in the early twentieth century. The telephone system consisted of effective monopolies, with the American Telephone and Telegraph Company controlling large networks of phones and phone lines. Customers did not even own their telephones but rather leased them from AT&T, or what was known as the Bell System, and nothing but the phone company's instruments could be attached to their lines. In the 1930s AT&T decided not to pursue the development of wirephoto or other facsimile transmission services over its lines, and since AT&T was a monopoly, this natural infrastructure for transmitting data was not readily available to independent inventors or engineers or to other companies. This is why newspapers used UHF radio waves as a transmitting medium in their earlier experiment. In the late 1960s the Federal Communications Commission, which regulated telephone networks in the United States, decided to allow non-Bell system equipment to use the established public switched telephone network (known in the industry as PSTN). Similar deregulation occurred in the early 1970s in Japan, which meant that facsimile transmission now had an intercontinental communication infrastructure available for its use. Hence, electronics companies and their engineers began to work with renewed interest on the development of improved fax machines that could be linked through the PSTN via acoustic couplers over which analog data was transmitted much as the voice was.

Competition had its downside, however. Growing numbers of electronics companies, driven by free access to the telephone network, soon introduced a plethora of new equipment that was not readily compatible with the other equipment on the market. This was clearly not a desirable development, for businesses who wished to invest in fax machines wanted to be able to send faxes to every other business with a fax machine, and yet communication between any two independently purchased units was not unlike a return to the earliest fax technology in which machines that were not paired in design did not work.

The dissemination of technology in a free market cannot proceed very effectively if each manufacturer works in total disregard of all the others. Purchasers and users like to be able to mix and match products and components for maximum convenience. When a new technology is being introduced, however, it is not uncommon for different manufacturers to be pursuing similar but different ways of dealing with the technical details. The first light bulbs, for example, did not have standardized bases, but now we expect different brands to be made with the same screw base so that they can be used interchangeably. Likewise, we can buy pencil leads by diameter and not worry about whether 0.5 mm Scripto leads will fit into a Pentel mechanical pencil, and vice versa. Such developments come about because manufacturers of like products agree to manufacture standard sizes and meet common performance criteria. When such agreements

On the numberpad of virtually every hand-held calculator and computer keyboard, the higher numbers are placed above the lower ones, while on the keypad of a touch-tone telephone the lower numbers are at the top. What could account for these two different arrangements? Do users who alternate frequently between calculators and telephones find the difference to be a serious problem?

Calculator	Telephone
7 8 9	1 2 3
4 5 6	4 5 6
1 2 3	7 8 9
0 · =	* 0 #

To make matters worse, the bottom line of different calculators can have various other arrangements or even different keys, and some telephones outside the United States have the * and # keys reversed. Is it desirable or practical to have the arrangement of all number keypads standardized?

do not arise voluntarily, government regulatory agencies might get involved in developing what are known as standards.

Unlike problems in mathematics, which practically always have a unique answer, a single problem in engineering and technology can have many different solutions. Thus, there are many different ways of providing portable power for electronic devices, and hence we have a plethora of batteries with different physical sizes and configurations, not to mention voltage output. Though this can be a great annoyance when trying to buy replacement batteries for a portable calculator, it is unlikely that the situation could easily be remedied, for it is unreasonable to require the various successful calculator companies to redesign their products around a single battery. While stores that sell batteries would certainly wish there were fewer kinds to stock, the large numbers that have evolved do take up relatively little space. And so the problem of having to deal with so many batteries is more of an annoyance than a big issue.

Video tapes of movies, by contrast, are more expensive than batteries, and they take up a relatively large amount of space to stock. In the early 1980s, two separate types of video cassette recorder (VCR) were competing for control of the market: Beta and VHS. The cassette used in Beta VCRs was over an inch shorter in length than the VHS cassette, with other dimensions smaller also, which meant that the VHS cassette was about 30 percent larger by volume. There was general agreement at the time that Beta technology could give better picture quality, but Sony did not license anyone else to use its patented Beta technology and thus more VHS recorders were sold by more different companies. In time it became much easier to find the VHS version of movies and other tapes in video rental stores, and it became the standard. Consumers also wanted to be able to exchange tapes with one another, and generally they did not want to have to deal with questions of compatibility. The less popular Beta was essentially driven out of the marketplace, and by the late 1980s a videotape cassette and a VCR in which to insert it essentially meant VHS technology, at least in the United States.

STANDARDS FOR FACSIMILE TRANSMISSION

Fax machines have an obvious need to communicate with one another. A given corporation could specify that all of its offices use the same brand of fax so that communications could flow smoothly among them, but communication across company lines would be impeded if fax machines were not compatible. Such compatibility would come if the different fax manufacturers agreed to common specifications, and the first standards for fax transmission were in place as early as 1968. Since communicating by fax, like telephoning, was an international issue, the United Nations, through its International Telegraph and Telephone Consultative Committee (known by the letters CCITT, because the official name in French is Comité Consultatif International Télégraphique et Téléphonique), promulgated the standards, which specified an analog frequency-modulation transmission scheme. According to the 1968 standard, an 8 1/2 by 11-inch page of information was expected to be transmitted within six minutes with a resolution of 100 dots per inch both horizontally and vertically.

While technical standards spell out minimum expectations at the time the standard is written, they do not limit the levels of performance that manufacturers can seek. All technological artifacts evolve as new ways are discovered and developed for removing limitations on existing technology. In the case of fax machines, some obvious limitations were the speed and resolution with which faxes could be sent, not to mention the high cost of some of the early machines. As engineers continued to work on such problems to give their company's fax machines an advantage over the competition, the 1968 standards became too modest. The class of fax machines that met only those standards came to be known as Group 1 machines. In 1976 Group 2 machines were defined as those meeting a more severe standard, one in which analog transmission time at the same 100 by 100 resolution was only three minutes.

American manufacturers of analog fax machines were not totally pleased with their sales in the late 1970s. They did not expect that a great deal of capital investment in a new technology would pay off very well, and so they chose not to pursue the development of digital fax technology the way the Japanese did. The situation was not unlike the way some American pencil manufacturers deliberately chose not to pursue the development of fine-line mechanical pencil technology, although in that case it was not just a question of return on investment, for it was feared that the newer product might compete with the established wood-cased pencil market. Thus, Japanese firms developed the American-invented technology, and eventually virtually all fine-line mechanical pencils sold in America were made in Japan. Paradoxically, though the new Japanese-made mechanical pencils were tremendously popular, their adoption did not significantly diminish wood-cased pencil sales.

In the case of fax machines, the Japanese were more motivated than

Americans to develop digital technology not just for marketing but for cultural reasons, because the multiplicity of phonetic symbols and ideographic characters in Japanese were not easily coded for transmission via telegraph and telex systems. Among the first to see this potential in the Japanese and Chinese markets had been Thomas Edison, and it had motivated his own early research on the fax machine.

By the 1970s the world's largest facsimile research laboratory was well established under the auspices of Japan's national telephone company. To promote a national fax industry, the Japanese government forced manufacturers to adopt a communications standard that was common for domestic and international telephone operations, funded research, and purchased the resulting products. By 1980 the CCITT put forth standards for Group 3 fax machines, in which digital encoding compresses the data to be transmitted, thus bringing the typical transmission time for a page to under a minute. The new standard resolution became 200 dots per inch horizontally and 100 vertically, with a fine resolution option of 200 by 200 dots per inch. Details regarding session protocol, known as "handshaking," which establishes how the data are to be transmitted between two machines, were also included in the new standard, and by the late 1980s convenient digital fax technology was well established and growing at an enormous rate.

Not only were Group 3 fax machines better at universal communication but they also were easier and more pleasant to use. Digital and computer chip technology had automated much of the work of establishing the communication protocol between sending and receiving faxes, and thermal printing enabled a relatively inexpensive and odor-free paper to replace the expensive and smelly paper that were common in analog machines. These technological developments made fax machines, which had frequently been relegated to mailrooms and operated by designated employees, more acceptable and used out in the open by everyone in offices everywhere. Indeed, since the newer fax machines, especially the more expensive office models that used plain paper, employed so much technology that was similar to that in a copier machine, in a pinch fax machines could double as convenient copiers.

SOCIAL AND CULTURAL FACTORS

Among other significant factors that accelerated the acceptance and use of fax machines were some that might be termed extratechnological or social. By the mid-1980s, thanks in part to some very effective television commercials and advertising campaigns, Federal Express had grown into a ubiquitous overnight delivery service. Indeed, the term Fedex came to be used almost generically for all speedy delivery services. In 1984, before every office had its own fax machine, the Federal Express organization invested heavily in a faxing service it called ZapMail, whereby faxed copies of documents could be delivered at unheard of speeds. Ironically, Federal Express's promotion of fax technology was so effective in selling the idea of faxing documents that the biggest potential customers bought their own fax machines and made the service unnecessary. Federal Express lost over $300 million in the venture, but its enormous success in overnight delivery enabled it to survive even such a large loss.

Another phenomenon of the 1980s that is widely believed to have influenced the rapid adoption of fax technology was the growing dissatisfaction with postal services. People were complaining more and more about how conventional mail (now derisively called "snail mail" by ardent e-mail users) was getting slower and less dependable. Letters were said to be lost and never delivered. At the same time, the presence and use of photocopying machines in offices had made office workers increasingly familiar with the concept of feeding documents into a copier and pressing a few buttons to begin the process. Many newer fax machines were remarkably like office copiers in their looks and operation, and so they were less intimidating to use and there was less resistance to their introduction into the office routine. Faxing a letter was just like long-distance photocopying. By 1987, for the first time, fax machines exported from Japan exceeded that country's domestic consumption of them. Furthermore, over the course of about a decade, from 1980 to 1992, the cost of digital fax machines dropped by a factor of 30, thus making them affordable by the smallest of businesses and even by individuals.

Over 600 different varieties of fax machines were available to the consumer by the early 1990s. In the United States alone the market for fax machines grew from half a million in 1985 to 6 million by 1991.

By the mid 1990s, personal computers were commonly being equipped with fax boards, which made it unnecessary to print out a computer-generated document before faxing it to someone who might then have to enter its data into another computer. Computer-to-computer faxes thus promised to reduce the amount of paper generated in offices, something the personal computer itself and the fax machine had not succeeded in doing. Estimates are that the number of pages transmitted via fax rose from 1.5 billion in 1985 to 17 billion in 1991, an increase of an order of magnitude in only six years. As much as 40 percent of telephone traffic between the United States and Japan was estimated to involve fax transmissions.

FURTHER DEVELOPMENTS IN FAX MACHINES

Both mechanical and electrical engineers, working individually and in groups on various aspects of the machine, played important roles in the development of the fax from a clumsy, slow analog device that used smelly paper to the compact, fast, and user-friendly machines that became as familiar as copier machines and telephones. In the early 1990s, for example, a group of Hitachi researchers and engineers wrote a paper on "compacting technologies for small size personal facsimile," which appeared in the *Transactions on Consumer Electronics* of the Institute of Electrical and Electronics Engineers. The six authors, led by a senior researcher whose name, Toyota Honda, suggested more an automotive than a communications engineer, came from five different divisions of the company: Hitachi Research Laboratory, Image and Media System Laboratory, Information System Equipment Operation, Telecommunication Division, and Mechanical Engineering Research Laboratory. The paper reported on two developments that were found necessary to produce a "small size personal machine," one involving rollers, which is a clearly mechanical engineering problem, and the other involving a computer chip, a clearly electrical engineering problem. Since home faxes are also expected to double as copiers, the quality of the image produced was very important.

Because a personal or home fax machine had to be inexpensive and small and yet produce a high-quality image, the mechanical engineering problem focused upon reducing the number of rollers and motors employed without significantly sacrificing printing quality. Larger machines contained two motors, one to drive the mechanism that handled a document to be transmitted or copied, and another to operate the printing end of the machine for documents received and for producing copies (see Fig. 6.2). The first motor drove three rollers, two of which served only to feed the document to the image sensor, which was located opposite a platen roller that played a much more central role in the process. The compact fax machine developed by the Hitachi engineers used only one roller to both feed and back up the image sensor, and a single motor was employed to drive both the input and output functions. Among the incidental problems that had to be addressed was the fact that a single roller with a

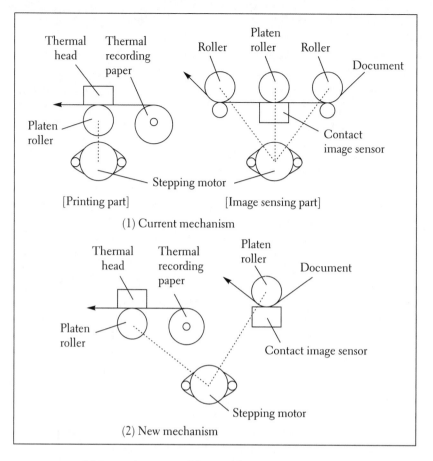

Thermal head

Thermal recording paper

Roller

Platen roller

Roller

Document

Platen roller

Contact image sensor

Stepping motor

[Printing part]

[Image sensing part]

(1) Current mechanism

Thermal head

Thermal recording paper

Platen roller

Document

Platen roller

Contact image sensor

Stepping motor

(2) New mechanism

FIGURE 6.2 Motor use in a personal fax machine

surface rough enough to grab and feed a piece of paper also tended to collect dirt. The new single-roller arrangement was thus dirtier, which translated into image distortion. This required the development of a new electronic means of compensating for shading on the document being read, and mechanical and electrical engineers had to interact on its solution.

Not surprisingly, the research paper of the Hitachi engineers does not reveal many of the details of how they ultimately solved their problems, but this is to be expected in an industry in which success depends so much on having a competitive product advantage. Industrial secrets cannot last long or be effective indefinitely, however. As soon as the new Hitachi compact fax would be distributed, competing companies could buy it as easily as consumers could. Engineers learn a lot about different ways of solving problems by taking things apart, an activity that many engineering students engaged in as children and many engineers continue to do throughout their lives. Trying to make a new machine by taking apart something of the competition's is known as reverse engineering, and it would have been immediately obvious that a big advance in the new Hitachi fax was its reduced number of motors and rollers. Understanding how the dirt problem was solved would take a little more work, but in time engineers working for competitors could figure it out, or at least understand the problem well enough to come up with an alternative and maybe even better solution. (If something is patented, of course, it cannot just be copied.) It is in such ways that competitive consumer products come to evolve into similar yet different forms. Because manufacturing companies know that their products are never safe from reverse engineering scrutiny—nor can they maintain their competitive advantage for very long—research, development, and engineering groups are an essential component of an active industry.

As the size and price of fax machines continued to be reduced, while reliability and quality of image seemed to be constantly improving, the question of speed of transmission remained an issue, at least with some manufacturers and users. Increasingly sophisticated encoding schemes required less data to be transmitted and hence decreased the time it took to scan and send a page, but the limiting factor was the 9600 bits per second that could be carried over the public telephone networks. In fact, fax technology had become so sophisticated that when a machine detected that it had gotten a noisy telephone connection, successively lower modem rates of 7200, 4800, and 2400 bits per second were automatically employed until one of these fallback speeds proved reliable.

The question of transmission speed led to a new standard being introduced in 1984, and faxes that fall within it are known as Group 4. In this category, fax machines are designed to communicate over the integrated-services digital network known as ISDN (see Fig. 6.3). However, ISDN itself did not become as widely available as soon as originally expected, due largely to the fact that local telephone networks and switching exchanges were not yet upgraded to ISDN capability. A decade later there were still relatively few Group 4 faxes operating. As with virtually all new technologies or technological advances, the situation attracted the attention of skeptics and nay-sayers and encouraged them to quip that ISDN stood for "It Still Does Nothing" and "I Still Don't Need." Whether the

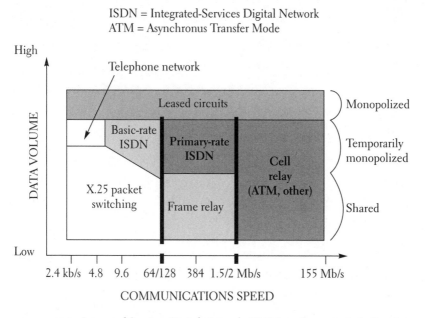

ISDN = Integrated-Services Digital Network
ATM = Asynchronus Transfer Mode

FIGURE 6.3 Integrated-Services Digital Network (ISDN) in the context of other data transmission technologies

letters might ever stand for "I Sure Do Now" depends upon what other new technologies might be developed before the ISDN network becomes widespread and before the costs of Group 4 fax machines drop enough to make them fully competitive. By the mid-1990s there were promising developments that ISDN would indeed become the "Interface Subscribers Definitely Need."

What new alternative technologies might be is not merely a technical question, as the case study of the fax itself makes clear. Even the most technically advanced problems have nontechnical components that affect their solution. Cultural, social, economic, and political developments can be limiting factors as much as the physical laws that govern electronic circuits and mechanical movements.

7

Like a lot of inventions, the tubojet airplane engine was developed more or less simultaneously and independently by different individuals in different parts of the world. In England, a young student named Frank Whittle became obsessed with the idea of employing jet propulsion in aircraft and wrote his graduate thesis on future developments in aircraft design. In 1930 Whittle applied for his first patent on the turbojet engine. In Germany, a few years later, another young researcher, Hans von Ohain, was encouraged by one of his professors to pursue interests in the jet engine. Whittle and von Ohain, initially with very little interest or support from their governments or military interests, led the development efforts in their respective countries, and jet fighter aircraft finally began to appear in the skies just as World War II was coming to a close. In 1991, for their determined efforts that ultimately changed the nature of aircraft worldwide, Whittle and von Ohain were named the second recipients of the prestigious Draper Prize by the National Academy of Engineering. (The first Draper Prize, awarded in 1989, went to Jack S. Kilby and Robert N. Noyce for their independent roles in inventing and developing the integrated circuit, which has revolutionized so much in the world, including air travel.)

Postwar Germany was in no position to develop a commercial jet airplane, but the British were and did—for use in such promising, lucrative, and revolutionary applications as economical long-range air travel, including transatlantic flights. The first commercial jet airline service, inaugurated on May 2, 1953, employed the jet-powered airplane known as the Comet, which was developed by the British de Havilland Aircraft Company. That manufacturer's distinct advantage in capturing the world aircraft market was short-lived, however, for exactly one year later a Comet exploded in midair while taking off from Calcutta. Within another year two other Comets suffered fatal midair failures, and the plane's design

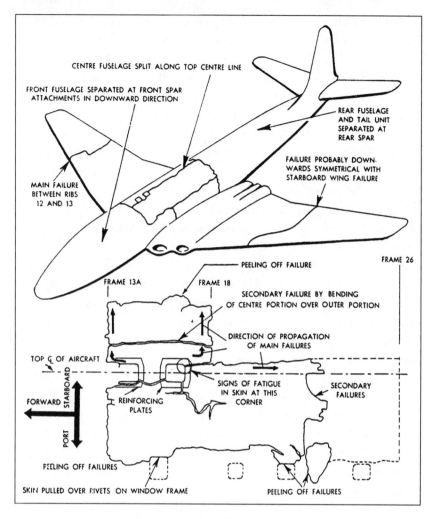

FIGURE 7.1 Reconstruction of a failed de Havilland Comet

came under serious investigation. Pilot error and bad weather were at first blamed for the accidents, but eventually the cause was identified as metal fatigue. Small cracks that developed in the highly-stressed corners of an opening in the plane's fuselage grew slowly but surely with each cycle from take-off to landing until they were so large that, when pressurized on the fatal flights, the plane shattered explosively and without warning (see Fig. 7.1).

Metal fatigue was a new phenomenon in aircraft. Before the introduction of jets, airplanes did not fly so high and thus did not have to be so highly pressurized for passenger comfort. In order to gain the fuel efficiency that gave the new engine part of its advantage, jets had to fly higher than propeller-driven airplanes, and as they did so the structural components of the aircraft were subjected to conditions that were beyond the experience of its designers. Previously, metal fatigue was believed to affect only machine parts that were subjected to cycles numbering in the hundreds of thousands, if not millions. Therefore, the Comet's engineers did not believe that fatigue would affect the plane, which would be subjected to far fewer cycles during its lifetime. But because airplanes must be as light as possible, their structural parts must carry more intense loads than the parts of land-based structures or machines, for which weight is less important. The critical combination of load intensity and flight cycles that would lead to the growth of critical cracks proved to be far lower than expected in the Comet.

The stresses, or intensity of force, to which a pressurized aircraft fuselage is subjected are not unlike those that an aluminum beverage can must resist. Thus, as a first approximation, the main body of the airplane may be thought of as a cylindrical pressure vessel. Since aluminum beverage cans generally are not subjected to widely varying cycles of pressurization and depressurization, fatigue is not a problem to be considered in beverage can design (though the phenomenon of metal fatigue can be easily demonstrated on an aluminum can by bending the top opening device back and forth several times). However, the airplane fuselage—which alternates between no net pressure difference between inside and outside when it is on the ground with its doors open to a large net internal pressure when it is flying at altitude in a rarified atmosphere— is subject to fatigue. Its metal skin can withstand only so many cycles of alternating stresses created by this repeated change of net pressure, the way the aluminum beverage can tab is able to withstand only so many cycles of being bent back and forth.

Because it took some time even to suspect that metal fatigue was causing the Comet crashes, and more time to correct the design flaw, the plane lost the confidence of the flying public, as did the airlines that flew it. As the British aircraft industry lost the advantage of having the first commercial jet airplane, other aircraft companies, especially in the United States, including Boeing, Lockheed, McDonnell, and Douglas (the latter two of which merged in 1967 to form the McDonnell-Douglas Corporation), began to develop their own models of jet aircraft. Their engineers had the advantage of learning from the failure of the Comet that metal fatigue was a major design consideration, and thus in the development of their own planes they could take special care to avoid the conditions of dangerously high alternating stresses in the fuselage.

Over the years, the Boeing Company came to dominate the world market for commercial jet aircraft, based largely on the successful development of its highly reliable 707 airplane. As air travel increased worldwide and as fuel prices became an increasingly large factor in the cost of operating an airplane, a variety of competing commercial jet aircraft were developed that provided a range of passenger capacity along with more fuel-efficient operating characteristics. Among these were the familiar DC-9 and Boeing 727 and, later, the wide-bodied aircraft such as the DC-10, L-1011, and Boeing 747.

Yet as early as the mid-1970s the American-dominated commercial aircraft industry began to be challenged by the European-based Airbus Industrie, a consortium of companies highly subsidized by the British, French, German, and Spanish governments. By the mid-1980s Airbus had begun to make significant headway into the world commercial airplane market, and in order to keep competitive in such a market, an aircraft company had to look constantly to the future to anticipate what kinds of planes would be desired by airlines as patterns of air travel would change with a changing world economy. In the late twentieth century, this meant paying special attention to what was happening in developing countries, especially those around the Pacific Rim. It was in this environment that, in the mid- to late-1980s Boeing began to look to designing and developing a new airplane that would fill needs that planes like its 747 and 767 did not (see Fig. 7.2). The new airplane design they decided to pursue became known as the 777.

CONCEPTUAL DESIGN

The design of a new airplane, like the design of anything else, begins at the conceptual or schematic stage; for until certain basic ideas are decided upon and put on paper or the computer screen, there can be little to discuss among the team members whose job is to carry out the detailed design that will result in a real object. Boeing's interest in developing a new airplane was at first market-driven: the company wanted to create a product that, in size, would fill a gap between the 747 and the DC-10 and L-1011 of the competition—aircraft that were aging and would eventually have to be replaced. Since it was an established fact in the aircraft industry that a new airplane took at least four years and a large financial investment to design, develop, assemble, test, and produce for sale, getting

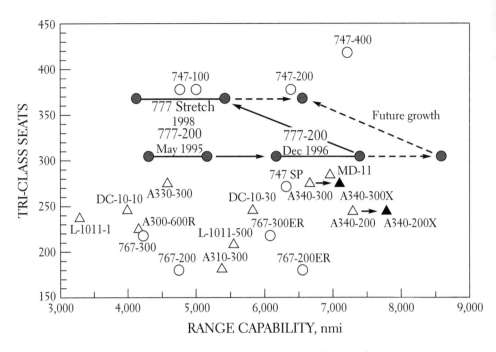

FIGURE 7.2 Chart comparing aircraft seating capacity and range of various commercial jets

started well before airlines would have to make replacement decisions was good business.

In the 1950s an airplane like the 707 cost about $15 million to develop and test; by the mid-1980s designing and developing a totally new airplane was expected to take billions of dollars, thus risking the very life of the company on the venture. At first Boeing looked to stretching its existing 767 design as a safe, quick, and low-cost way of providing a plane with increased seating capacity. Developing a stretch version of an existing aircraft is relatively easy, in that sections of fuselage are added before and after the wing, thus altering in a minimal way the location of the center of gravity and the aerodynamics of the basic design. The new airplane was referred to as a 767-X, connoting the fact that it was a version of the then-familiar and reliable 767, which had a capacity of about 200 passengers, more or less, depending upon its configuration. In the meantime, however, McDonnell-Douglas was introducing its new 323-seat MD-11, which was a modernized DC-10 with fuel-efficient engines and was expected to be available in 1990. In addition, Airbus was expected to be delivering its new 300-seat four-engine A-340 in 1992, and a year later the slightly larger capacity but twin-engined A-330 model, which would also have a longer range.

Boeing came to believe that, with such competition on the horizon, the better marketing strategy would be to develop an airplane that could compete more directly with other new products. Thus, in late 1988 the company decided that it would offer an entirely new airplane, the 777. Conceiving of a new airplane, on the one hand, and committing the corporation to designing and developing it, on the other, were actions that were about $4 billion apart. In the aircraft industry a firm commitment to go ahead with detailed design development does not take place until there is a commitment by a customer to buy a certain number of the new planes. Almost two years elapsed before United Airlines became Boeing's "launch customer," which meant that the manufacturer could announce its firm multibillion-dollar financial commitment to going ahead full speed with the project. Thus the Boeing 777 was launched in late 1990, into a market in which dozens of airlines had already ordered hundreds of MD-11s, A-340s, and A-330s.

As part of a marketing strategy to catch up with the competition, Boeing invited eight American and overseas airlines to get involved with the design of the plane in its early conceptual design stage, when few things were firmly decided upon and so changes were easily made to suit the customer. The airlines included, besides the launch customer United, All-Nippon, American, British, Cathay Pacific, Delta, Japan, and Qantas, and they met with Boeing engineers over the course of a year to express their desires. While no one expected such a diverse group of customers (which had come to be known collectively as the "Gang of Eight") to agree on what would be the ideal airplane, enough points of consensus were reached to fix a design concept that could serve as the basis for detailed engineering work. Among the clearest points of agreement was that the width of the fuselage should be greater than that of either the MD-11 or the new Airbus models.

The size of an airplane's fuselage, like the span of a bridge, is the kind of conceptual design decision that has far-reaching technical and cost implications for the project. Only after such dimensions are set can detailed and interrelated calculations involving strength, weight, aerodynamic properties, power requirements, and the like be made. For example, the stress developed in the fuselage structure that holds each section of pressurized cabin together is directly related to the cabin's diameter, as is the drag the aircraft will experience in flight. Greater stress requirements mean greater amounts or stronger structural materials must be used, which translates into more weight or cost. A heavier plane requires more power to move, which translates into more total engine thrust. Alternatively, of course, the passenger capacity of the plane can be reduced, but this goes against the primary reason to have a wider cabin, namely, to carry more passengers. In short, the decision of how wide an airplane's cabin will be must be fixed early in the design process—at the conceptual design stage—because so much else about the design depends upon that single dimension.

Paradoxically, by starting behind the competition, Boeing could have an advantage over it. Since potential customers found fault with the cabin widths of the planes with which Boeing wished to compete, it decided to make the inside cabin width at armrest level about five inches wider than that in the MD-11 and about twenty-five inches wider than that in the A-330 and A-340. This translated into a flexibility of seating arrangements that could mean as many as 30 additional seats and paying passengers per plane. Another problem that airlines found with a lot of existing aircraft was the design and arrangement of overhead storage bins. When Boeing presented a design for overhead compartments that increased headroom when they were closed but opened in such a way that even shorter passengers could gain access, the Gang of Eight were unanimously in favor of incorporating the bins into the 777, even though some of the engineers argued that the bins were harder to manufacture and added weight to the aircraft. Such tradeoffs are always encountered in engineering design, however, and what is the "best" design decision is not always determined by technical considerations alone; in this case passenger satisfaction was assigned a high priority.

American Airlines proposed a more far-reaching design feature, prompted by the fact that the required wingspan of the 777 was estimated to be roughly 200 feet, or about 45 feet longer than that of the DC-10. This would mean that the 777 would not fit into existing airport gates without sacrificing capacity or flexibility. Since airports were already crowded, giving up gate space was not an attractive option, and the wider wingspan would also present problems on the relatively narrow taxiways of some older airports. American proposed that the 777 have wing tips that folded up for taxiing and gate parking (Fig. 7.3). This feature, though

FIGURE 7.3 Folding-wing design of Boeing 777

familiar on military aircraft in order to conserve space on aircraft carriers, was new for commercial planes and was not without its drawbacks. On smaller fighter aircraft, there can be visual and manual inspection by the ground crew that the wing-locking mechanism was firmly engaged before takeoff. The size of a 777 would make such inspection difficult, and so elaborate safety features would have to be incorporated into the design and considerable precautions would also have to be taken to ensure that the wing-locking device would not accidently become unlocked while the plane was in flight. Furthermore, the structure and machinery to make the wings fold and lock would add considerable weight to the plane. Nevertheless, Boeing agreed to incorporate the folding-wing feature into the design so as not to lose future business owing to an inability of the 777 to fit into the existing air transportation infrastructure.

TRADITIONAL AIRCRAFT DESIGN

The traditional way of carrying out the detailed design of a new aircraft was for a lot of engineers and draftspersons to work individually and in teams on various parts and subsystems of the plane. In the 777 there were over 130,000 unique individual parts to be engineered and, when rivets and other fasteners are counted, over 3 million total parts to be assembled into each plane. The 747, which had a total of 4.5 million parts, required about 75,000 individual engineering drawings to specify. Of course, this great number of drawings all had to be consistent with one another if the parts were to fit together and if the various cables, wires, and ducts were not to interfere with one another. To check that parts and systems were compatible, an engineer working on one part had to get the drawings of mating parts, and any change that might have to be made in one area necessarily had sometimes fundamental implications for other parts or systems. It was a slow, arduous, and frustrating process, and one senior draftsperson at Boeing recalled "waiting for days to get someone else's drawing, getting a copy made, and slipping it under mine to trace their part." Even with careful checking and cross-checking, interferences and mismatches frequently occurred.

To catch incompatibilities among systems, physical mockups were constructed in which all the tubing, wiring, ductwork, and the like were installed. If a tube had to pass through a duct, for example, a rerouting would have to be devised, which meant that all of the associated drawings would have to be revised as well. Physical mockups were expensive, labor-intensive, and time-consuming, and they added to the cost of the airplane.

Another big problem associated with coordinating a great number of individual parts was that when it came time finally to assemble the actual airplanes, not everything fit quite exactly. Shims of one kind or another had to be inserted to smooth out the poor match of fuselage parts, even though they had been manufactured according to specifications. For example, the 747 contained about 1000 pounds of shims, which not only added to the cost and labor of the plane but also added unplanned-for dead weight. Boeing took extra care in the design of its 767 to reduce, if not eliminate, the necessity of shimming.

The cost of additional change orders for parts after they have been specified by presumably "final" drawings is among the more difficult expenses to plan for in design and manufacturing. Furthermore, such changes add no new value to a product, and so their cost cannot be recovered by the manufacturer once a price for the aircraft has been set. Management approaches that go under such names as Total Quality Management (TQM) and, at Boeing, Continuous Quality Improvement (CQI) have aimed to avoid these costly mistakes by insisting on a high level of cooperation and effort throughout the entire design process.

To design the 777 as efficiently and cost-effectively as possible, while at the same time achieving high quality, Boeing chose a "paperless" design strategy. This meant that computers were to be used in the design, testing, and manufacturing process to a greater extent than ever before. Boeing's earlier limited experience with computer-aided design (CAD) consisted of designing an engine strut for the 767. That design was completed $3\frac{1}{2}$ months ahead of an allocated 24 months, and at a cost that was lower than estimated for then-conventional design methods. In addition, the number of expensive changes to the design subsequent to its being released was significantly lower than for other Boeing strut designs. The CAD system used for this impressive pilot program was a Computer Aided Three-dimensional Interactive Application (CATIA) developed by Dassault Systemes, a French-owned software development firm with close ties to IBM.

Scaling up the CATIA system to enable it to handle the volume of data and number of users required to design the 777 entirely by computer was in itself a computer-engineering design problem of unprecedented magnitude. The installation at Boeing's Everett, Washington, factory, the principal location for 777 design and assembly, was to have over 2000 terminals (see Fig. 7.4) connected to eight of IBM's largest mainframe computers. At first the CATIA system was operated on ES/9000-720 machines, but they were replaced with larger capacity ES/9000-900 units before the peak design demand that was expected to occur in 1992. Image manipulation, which is so important in CAD but which consumes so much computer memory, was done locally on IBM 5085 and 5086 workstations. The total storage capacity required for the overall system reached a staggering 3.5 terabytes, which, if stored on high-density 3.5-inch 1.44-megabyte disks, would require about 2.5 million of them, or a pile almost five miles high.

As many as 238 teams, comprising as many as 40 engineers, were involved in the design, development, and manufacturing of the 777, and they all needed access to all of the computer data. A truly paperless design meant that, instead of waiting for physical drawings to be copied and checked for compatibility of parts and systems, an engineer working on a single part could call up all mating parts or systems into which it was to fit on any one of the over 7000 workstations that were eventually spread around the world in over 17 time zones. Among the steps taken by Boeing

FIGURE 7.4 A CAD workstation used in designing the Boeing 777

to keep this data network efficient, accurate, and secure was the laying of dedicated data lines across the Pacific Ocean from Washington to Japan. There, the Japanese Aerospace Industry, a consortium of Fuji, Kawasaki, and Mitsubishi Heavy Industries, was responsible for about 20 percent of the fuselage structure. Involving the Japanese was a way not only to share the investment cost but also to promote airplane purchases by Pacific Rim airlines.

The software of the computer-based 777 project enabled even the most subtle of interferences between parts or systems to be identified via an electronic preassembly program known as CLASH. The interferences were highlighted by flashing red zones and the identity of the parts and the names of the engineers responsible for them were available on the screen for immediate contact and attention. To be sure that the newly developed CAD system was itself reliable, early on in the design process some conventional physical mockups of aircraft subsystems were constructed. The computer system proved to be so effective at pinpointing interferences (see Fig. 7.5) that further physical mockups were not used.

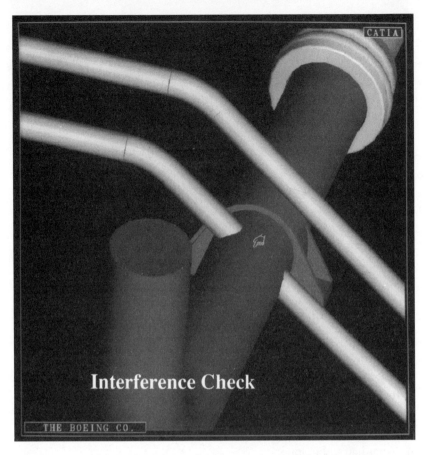

FIGURE 7.5 Example of interference between systems caught by computer

As for the accuracy of fit, the CAD program, in conjunction with computer-aided manufacturing (CAM), in which digital data were fed directly into computer-controlled fabrication machinery, produced a 20.33-foot diameter fuselage whose components aligned within 0.023 inch vertically and 0.011 inch horizontally along the length of the airplane body—better than one in ten thousand. There was certainly little need for shims in such an assembly.

Before the advent of CAD and CAM and the ease of communication they have allowed between design and manufacturing engineers, in most technological enterprises the drawings for individual parts were said to be "thrown over the wall" separating the design from the manufacturing teams, with the sometimes overly optimistic expectation hope that the part could in fact be made and mated with other parts. There had been a few notable exceptions to this short-sighted practice, and one was the division of Lockheed known as the Skunk Works, where design engineers worked in the same building in which their highly classified designs were assembled. The Skunk Works was responsible for such supersecret and technically supersuccessful projects as the U-2 spy plane and the F-117A stealth fighter. By having engineers and manufacturers share data and ideas as early as possible in the design process, and by having them work together in teams so that each knows what the other is thinking and can do, it is believed that misfits and related problems were cut in the 777 to such a great extent from what was traditionally expected that for competitive reasons Boeing has not said exactly how efficient the process actually was.

In addition to producing parts that could physically be manufactured and actually mate with a minimum of shims, if any, the digital-computer-based design-through-manufacturing system introduced in the Boeing 777 eliminated another traditional glitch in the process. In the past, one of the big surprises that could arise even when parts fit perfectly was access to those parts, first by assembly workers and later, when the plane was in service, by maintenance workers. As a way of anticipating, at the design stage, situations where clearance was inadequate for a worker to reach a part of the plane, Boeing developed a simulated mechanic called CATIA-man (Fig. 7.6), who could be manipulated to crawl around inside the assembled digital plane to be sure that there was enough elbow room to do what had to be done.

FLY-BY-WIRE

The role of the computer was not limited to the design and production of the 777. This aircraft is Boeing's first airplane to have its flight control system operated electronically through twisted pairs of wires that connect the pilot's input to the aircraft's flight control output systems through the intermediary of on-board computers. All previous Boeing aircraft were controlled mechanically through a system of levers, cables, pulleys, and hydraulic lines that connected the pilot directly to the aircraft's control surfaces. Total "fly-by-wire" technology (as computerized flight control is called) was first employed in a commercial airliner by Airbus, when its A-320 was introduced in 1988. However, that airplane's fly-by-wire system came under scrutiny when some A-320s crashed in circumstances that called into question the reliability of the software that was at the heart of computer-controlled aircraft.

FIGURE 7.6 CATIA-man, developed to check human maneuverability during construction and maintenance operations

Among the criticisms of the Airbus approach to fly-by-wire was that the on-board computer would not allow pilots to carry out certain maneuvers that were deemed too extreme for the aircraft. Thus, through its software programs, the computer limited the amount of acceleration the pilot could call upon so that the plane's structure would not be overloaded. Early versions of the software allowed the aircraft to fly very close to the ground at very low speed, but did not allow enough time for the engines to recover full power in the event that it was needed to abort a landing or react to some unforeseen condition. This limitation was removed after the crash of two A-320s.

Another characteristic feature of the Airbus electronic control system was the relocation of the control stick from between the pilot's legs to a position beside the seat. But neither the new side-stick controllers nor the throttles moved when the computer automatically adjusted power to maintain flight conditions, and this left the pilot without tactile or visual feedback about exactly what the plane was doing at any given time.

Boeing's approach to fly-by-wire in the 777 kept the pilot more in touch with the feel of the aircraft. For example, in addition to interpreting the pilot's commands and sending them to the engines and control surfaces, the computer was programmed to move stick controls and throttles to reflect the plane's condition. As airspeed changed, a force would build up in the control column to give the pilot a sense of how much acceleration the plane was undergoing. As for overstressing the airplane, Boeing's philosophy was that if an emergency situation arises, such as an impending midair crash, the pilot should be able to turn, dive, and accelerate as quickly as required to avoid the collision, even if the maneuver stresses the plane to the point where it suffers some structural damage. Such differences in philosophy are characteristic of all kinds of technology, and understanding their implications for reliability and safety will become increasingly important as more and more powerful and complex computers and computer programs are developed.

In addition to the fly-by-wire features of the 777, Boeing introduced another innovation into the cockpit by replacing the cathode ray tube (CRT) displays with flat-panel full-color liquid crystal displays (LCD). As with technological change generally, there are pluses and minuses associated with the different displays. The CRTs installed in 747s, for example, were found by pilots to be hard to read in strong sunlight, and the LCDs do not have a strong image in cold weather, which means that for such conditions there will have to be an auxiliary heating system installed. On balance, however, the change is claimed to be for the better.

ENGINES AND ECONOMICS

Among the early selling points of the 777 was the fact that its two engines would be powerful and reliable enough to enable the plane to complete transoceanic flights even if one engine failed. Airlines were initially attracted to two-engine planes because, generally speaking, they use less fuel per passenger and require less maintenance than three- and four-engine models. But the mix of cost factors associated with operating an airline is not static, and by the time the aircraft was delivered other money-saving features had become selling points as well. Whereas the price of fuel was at one time the most important component of total cost, by the 1990s wages associated with maintenance and operation, as well as debt service, were the dominant factors. The fuel efficiency of two-engine aircraft was thus a bonus on top of the reduced cost of maintenance. Operating costs were less because the 777's fly-by-wire cockpit eliminated the need for a flight engineer, thus cutting the size of the flight crew by one-third. And the lower purchase price of a two-engine plane eliminated some of the financing costs for potential customers.

Since the 777 was the biggest twin-engine jetliner ever put into operation, it had to have the most powerful engines ever made. The first 777s were fitted with Pratt & Whitney engines designated PW4000, which were capable of producing a thrust in the range of 80,000 pounds. This was over one-third more thrust than the most powerful engines on 747 jumbojets. The 777 engines are described as high-bypass turbofans, which employ large, wide-bladed fans visible in the front of the engine to take in great amounts of air (Fig. 7.7). While some of the air is used to feed the turbine, which in turn rotates the fan, a considerable amount of the air bypasses the turbine and assists in propelling the plane.

To make the 777 more attractive to a wide variety of airlines, whose maintenance crews and shops would not have experience or tools to deal with Pratt & Whitney engines without additional investment, alternative engines were offered. Among the options was a General Electric engine, which is so large that its nacelle—the streamlined outer covering that gives an engine its visible shape—is about the same diameter as the fuselage of a Boeing 757 airliner. Another engine option available on the 777 was one made by Rolls Royce. Pratt & Whitney and General Electric engines are also options on Airbus planes.

The fact that the 777 was designed to fly extended distances with one engine out made it a candidate for immediate transoceanic travel, if it could secure the necessary flight certification. In the past certification came only after extensive experience once the aircraft was put into commercial service. In the mid-1980s, however, after it was variously estimated that the chance of two engines independently breaking down simultaneously was one in a billion hours of flight or once every 50,000 years, the Federal Aviation Administration (FAA) began to consider allowing twin-engine jets to fly not only transatlantic routes but also transpacific routes. Such developments were critical to the success of Boeing's 767 and 777.

Although Boeing considered the 777 to be an entirely new airplane model, the company emphasized that is was an evolutionary rather than a revolutionary development. The distinction was significant, for it meant

FIGURE 7.7 Pratt & Whitney engine mounted on a Boeing 777

that the plane could be expected to receive expeditious flight certification. Among the additional arguments for such consideration was the fact that the 777 was to be subjected to unprecedented testing via computer before it was even assembled. Furthermore, its Pratt & Whitney engines were not totally new designs but were evolved from earlier models that had already demonstrated considerable reliability and were to undergo extensive flight tests before the first airplane was delivered. Thus, years before the first 777 was handed over to United Airlines in 1995, Boeing anticipated that the FAA would look favorably upon allowing it to skip the customary two-year test period during which the new plane had to remain within 60 minutes' flying time of a major airport. This would, of course, have precluded transatlantic flights during the probationary period. The FAA waived that requirement as expected, and the inaugural flight of the launch customer United Airlines was from London's Heathrow to Washington's Dulles Airport in June 1995. By the end of the year, the 777 had proven to be 97.5 percent reliable in meeting schedule demands, which compared with 94 percent for the 747-400 when it was introduced. United Airlines, however, was looking for even better performance and before the 777 was a year old had written a letter of complaint to Boeing. The problems identified would send the engineers back to their electronic drawingboards.

HUMAN FACTORS

The field of human factors, also known as ergonomics, endeavors to make the interface between a person and a piece of technology as logical, safe, comfortable, and friendly as possible. In the past, engineers have often been accused of not worrying about, or being incapable of appreciating the importance of, how their products were actually used by real (and fallible) people. This was reflected in jokes told about everything from incomprehensible user's manuals to electronic devices whose controls and settings were impossible to master.

Consideration of human factors is critical in designing the cockpit of an airplane, of course, and for some time considerable attention to detail has been paid to this area of technology. This is why the dials, switches, levers, buttons, and other meters and controls have distinctly different looks and feels to emphasize their different meanings and functions. From mishaps in the nuclear power industry we know the dangers that can arise when designers ignore this consideration. Some erroneous operator actions in the control rooms of early nuclear power plants were attributable to the fact that more importance was paid to making the components on a control console match and look "modern" than to designing different controls, dials, and gauges with distinguishing features. Similarly, the 1992 crash of an A-320 into a mountain near Strasbourg, France, as it was on a landing approach was attributed to the pilot's apparent confusion about the descent mode chosen (see Fig. 7.8). Crash investigators speculated that a 3.3-degree flight path was desired, but instead a 3300-feet/minute descent mode was punched into the computer. When the cockpit screen displayed 33, indicating the descent mode in hundreds of feet/minute, the pilot mistook it for 3.3, and thus did not realize that the plane was

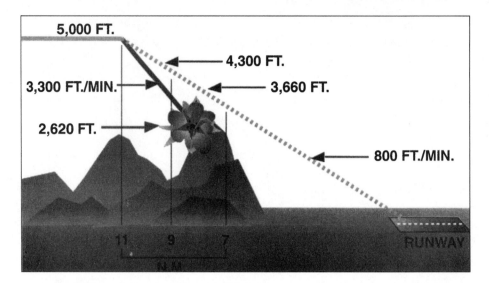

FIGURE 7.8 Airbus descent modes possibly confused in computer-controlled landing procedure

descending at a faster rate than it should. A display designed with human factors considerations in mind could have shown the angle of descent as two digits (3.3) and the altitude change as four digits (3300), thus making a confusion of the two very unlikely.

Perhaps because passengers, as opposed to pilots, were not seen to be directly interacting with a machine, designers of airliner cabins have traditionally paid more attention to aesthetics than to human factors. By realizing that, from the point of view of the airline, the passenger is the customer, and by deciding to include the airlines in the design process early on, Boeing was able to incorporate more passenger-friendly features into the cabin of the 777.

For example, in addition to the higher ceilings and overhead bins that at the same time swing down to be more accessible, the cabin is fitted with overhead lights whose burned-out bulbs can be changed quickly and easily by a flight attendant. Rather than having only one movie to view and having to crane their necks or sit on pillows to peer over the seat in front of them in order to see a centrally located movie screen, 777 passengers have individual entertainment monitors at their seats. Digital input received from a satellite communications network is transmitted to the individual monitors via fiber-optic cables throughout the aircraft. Each seat is also fitted with individual telephone sets whose keypads double as

The armrests of airplane seats were long manufactured with flip-top ashtrays, for the use of passengers who wished to light up in the air. In the early days of flight, a large proportion of the sophisticated flying public did smoke, but with first the decline of smoking, and then its outright ban on domestic flights, the ashtrays became an increasing annoyance for airlines. After every flight, the convenient little receptacles were stuffed with gum and peanut wrappers, and were time-consuming to empty. To put a stop to this practice, the airlines welded the ashtrays shut.

Was this solution a failure of imagination on the part of the airlines? Can you think of any alternative measures the airlines could have taken to covert the ashtrays into something useful?

an input device to the screen, thus enabling interactive video games to be played and information banks to be accessed. The case history of the 777, like that of virtually every large and complex technology project, shows that engineers do their job best when they interact with those who will be the customers of the product, whether they be the airlines or the passengers.

8

WATER AND SOCIETY

According to an early nineteenth-century definition, civil engineering is "the art of directing the great sources of power in nature for the use and convenience" of everyone. Later in that century a railroad engineer defined engineering generally as "the art of doing well for one dollar, what any bungler can do with two." These two ideas, social end and economy, are inextricably intertwined with virtually everything engineers do, although sometimes the relative ratio of costs to social benefits is less than self-evident. Many of the products of modern manufacturing, such as paper clips and aluminum cans, are clearly conveniences, and civilization did exist without them and would no doubt continue to do so with little significant change if they were banned. More substantial products of engineering, such as bridges and jet aircraft, are in a different category, and the nature of transportation as we know it today would be drastically changed if cities like San Francisco had to rely entirely on ferries to cross the bay or if we had to return to travel by ship to cross oceans. For one thing, the amount of time such crossings would take would increase by an order of magnitude. With a less-developed transportation network, there would be fewer imported products, northern cities would not have citrus fruits so readily in winter, tea and coffee would be dearer, shoes and clothing would be manufactured more locally, and so forth. Nevertheless, society could still function, albeit at a slower, nineteenth-century pace.

The economic use and control of water is in yet another category from both cans and airplanes because it is an essential ingredient of life itself. Like warmth and shelter, people have always needed water, and a considerable body of specialized engineering knowledge and techniques has developed over the centuries to provide, control, treat, and dispose of water for ever-growing communities of people. Individuals can survive in primitive conditions by living on insects and rainwater, as the Air Force pilot downed in 1995 in Bosnia did for almost a week, and we can imagine that small bands of our nomadic ancestors were able to live on similarly scarce resources. But we prefer broader diets. The development of agriculture, which might be described as land and crop engineering, was intertwined with the ability to move water from where it was to where it was needed when the weather was not cooperative. In the millennia since the establishment of agrarian societies, the control of water has been one of the great achievements of engineering.

WATER SUPPLY AND REMOVAL

Roman cities are famous for their water supply, which made possible their public fountains and baths. Some of the grandest surviving Roman arch bridges, such as the Pont du Gard near Nice in southern France (Fig. 8.1), were built for transporting not people but water. They are the aqueducts designed to carry great quantities of water from the hills to the city. Vitruvius, in the oldest surviving book of engineering, described the lead and clay pipes used in the Roman water supply system, cautioning against the former as a health hazard. He also explained how reservoirs should be constructed at strategic locations and how pressures in the pipes could be kept under control. Loss of water due to leakage and unauthorized removal from the Roman system was a chronic concern, and it has been estimated that half of the water entering the aqueduct system was lost before reaching its intended destination. There were thus plenty of problems presented for solution by Roman hydraulic engineers. Many of them were identified by Frontinus, the first-century Water Commissioner of Rome, who wrote in a famous report on the city's water supply that there was no comparison between the "idle Pyramids, or those other useless though renowned works of the Greeks" and the "indispensable structures" of the aqueducts.

At the time Frontinus was writing, in 97 A.D., Rome was being supplied with 40 million gallons of water per day, according to one estimate, thus providing its one million inhabitants with a per capita supply very close to twentieth-century standards. The vast majority of Romans did not use such great amounts of water individually, of course, for they carried their daily supply in vessels from public fountains. Great amounts of water were diverted to the public baths and to laundries where togas were washed. But among the main reasons that a steady and dependable source of water was needed by Rome and other ancient cities was that it had to be available at all times to fight fires and to keep the city clean. Such

FIGURE 8.1 The Pont du Gard, a critical link in a Roman aqueduct

overarching benefits to the whole community clearly justified the great effort and expense to design, build, and maintain a reliable water supply system in the first place.

Even with the abundance of water delivered to Rome, only the wealthiest or most influential or devious of its citizens had water piped directly into their homes, but there was no organized sewer system to remove it after use, nor would there be a significant one in any city until the nineteenth century. The disposal of waste water, and waste generally, was left to each individual and household to solve on its own. Systems of large underground sewers for storm and waste water were well established by Frontinus's time, but the only direct connections to them were constructed at private expense. Most citizens used public latrines or emptied pots of kitchen and toilet waste directly into the city's gutters, which were flushed regularly into the sewers. When the rain did not do so, large quantities of water were also used to flush the sewers themselves, lest the

The expansion of the Paris sewer system during the Second Empire and Third Republic was a technological and political triumph. During the Exposition of 1867, the sewer administration began offering public tours of "un second Paris souterrain." The public descended into the sewers by means of an elegant iron stairway and rode in deluxe versions of the vehicles used in sewer cleaning (see Fig. 8.2). Another part of the tour was made in carpeted and cushioned boats reminiscent of Venetian gondolas, each of which could hold sixteen people.

Most sewer systems throughout the world are not open to the public, but it is important to have access in order to be able to inspect, clean, and repair them. A well-designed sewer system will have access holes (manholes) located throughout its length. Why are the covers on these manholes almost invariably round?

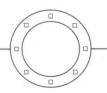

FIGURE 8.2 Sewermen walking wagons of visitors through the Paris sewers in the 1870s

odor and filth be unbearable. The importance of this function was spelled out explicitly by Frontinus: "I desire that nobody shall conduct away any excess water without having received my permission or that of my representatives, for it is necessary that a part of the supply flowing from the water castles shall be utilized not only for cleaning our city but also for flushing the sewers."

What to do with the contaminated water that was flushed from the sewers is a problem faced by cities to this day. Not until the mid-nineteenth century were civil engineers able to attack it in a systematic way. Engineers specializing in such problems became known as sanitary engineers, a name that was current until about the 1970s, when it was superseded by the term environmental engineers to reflect the profession's interest in an ever-growing list of problems dealing with waste and the pollution of the environment.

SANITARY SEWERS

The solution of large-scale engineering problems that affect diverse constituencies takes more than just technical know-how, for there has to be the societal resolve to impose the solution in an effective way and the financial means for doing so. Such conditions were in place in London in the mid-nineteenth century. (In earlier centuries, the death rate had often exceeded the birth rate, thus keeping down the growth rate of cities and obviating any urgent need to develop radically new technologies.) Water supplies had included the sixteenth-century scheme of pumping water from the River Thames to the higher parts of the metropolitan area, from which it could flow to its ultimate destinations by the force of gravity. Some pumping was achieved by means of power derived from the tides through waterwheels placed in the narrow archways of old London Bridge. Also in use were neighborhood pumps that raised local groundwater for community use. In time, it was recognized that these potential sources of contaminated water from the river and the ground, into which dirty water flowed, resulted in intolerable health risks, even if the exact nature of those risks was not fully understood.

Sanitary conditions in London in the early nineteenth century were abominable. There was a Commission of Sewers for the City of London proper, but its jurisdiction contained only about five percent of the 300,000 houses and of the total population of 2,500,000 in the growing metropolitan area. Regional sewer systems existed, but they were poorly designed and often ineffective, with large ones discharging into smaller ones, which naturally could not accommodate the full flow. After a comprehensive study of London sanitary conditions, one engineer reported in 1847 that there were houses without any drainage and with overflowing cesspools, and there were hundreds of streets without any sewers at all. At the time, an outbreak of cholera in India was working its way westward, and a royal commission advised that a single sewer district should be created to deal with the mess. Parliament, which met on the banks of the putrid Thames—then effectively an open sewer—agreed and passed legislation creating a Metropolitan Commission of Sewers, but little was done before more than 25,000 deaths due to cholera were recorded over the next six years.

In the meantime, Joseph W. Bazalgette was appointed chief engineer of the Metropolitan Commission of Sewers, and he described the situation as he found it:

> According to the system which it was sought to improve, the London main sewers fell into the valley of the Thames, and most of them, passing under the low grounds on the margin of the river before they reached it, discharged their contents into that river at or about the level, and at the time

only, of low water. As the tide rose, it closed the outlets and ponded back the sewage flowing from the high grounds; this accumulated in the low-lying portions of the sewers, where it remained stagnant in many cases for 18 out of every 24 hours. During that period, the heavier ingredients were deposited, and from day to day accumulated in the sewers; besides which, in times of heavy and long-continued rains, and more particularly when these occurred at the time of high water in the river, the closed sewers were unable to store the increased volume of sewage, which then rose through the house drains and flooded the basements of the houses. The effect upon the Thames of thus discharging the sewage into it at the time of low water, was most injurious, because not only was it carried by the rising tide up the river, to be brought back to London by the following ebb tide, there to mix with each day's fresh supply—the progress of many days' accumulation toward the sea being almost imperceptible—but the volume of the pure water in the river, being at that time at its minimum, rendered it quite incapable of diluting and disinfecting such vast masses of sewage.

Understanding and articulating the problems with the existing system—as Balzalgette clearly did in this passage, which may be described as a comprehensive failure analysis—is essential to working out an engineering solution to a problem. He designed a sewer system to obviate the failures of the existing system. The capacity of the system was established as being equal to the water supplied, and he designed the sewers to be large enough to carry that amount and to be laid at a sufficient slope to ensure that, even half full, the flow would be fast enough to prevent silting. Instead of emptying into the Thames in the middle of London, Bazalgette designed large intercepting sewers to run parallel to the banks of the river and to carry the waste water well downriver of the metropolitan area before discharging it. The sewer system was designed as an integral part of the Thames Embankment (Fig. 8.3), which serves to this day as a popular fresh-air promenade along the river.

For his pioneering engineering work that changed the London riverfront from an open sewer to a tourist attraction, Bazalgette was knighted in 1874 and is commemorated today on the central London area of the embankment known as the Victoria Embankment by a bronze plaque set in a monument at the foot of Northumberland Avenue, near Hungerford Bridge—within sight of monuments to the civil engineer Isambard Kingdom Brunel and the electrical engineer Michael Faraday.

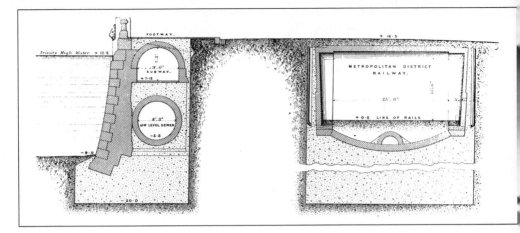

FIGURE 8.3 Cross-section of the Thames Embankment, illustrating how it conceals a sewer that carries waste downriver from London, in addition to an underground line

THE CASE OF CHICAGO

Since each city's geographical and topographical situation is unique, each presents unique engineering problems of water supply and disposal, and the solutions often have to be modified as a city grows. Mid-nineteenth century Chicago took its water from Lake Michigan and returned it there and to the Chicago River, which emptied into the lake. In order to gain a fresh supply of water in the 1860s, a five-foot-diameter tunnel was run for two miles under the bed of the lake, far enough for any sewage to have been sufficiently diluted. The water was pumped to the 154-foot castle-like tower that still stands near Water Tower Place, which was named for the historic location. From the Water Tower, distribution proceeded by gravity flow.

With the growth of the city, however, increased sewage polluted the lake for miles offshore, and a new system of dealing with contaminated water was sought. In the last decade of the nineteenth century, Chicago began dredging operations to reverse the flow of the Chicago River, so that it drained not into Lake Michigan but into the Des Plaines River, which in turn flowed into the Illinois River and, eventually, into the Mississippi. While Chicago had thus protected its lake water supply from its own sewage, there was some question as to whether the diluting and oxidizing action of the rivers would cleanse the water sufficiently before it reached those communities downstream that relied upon it. The state of Missouri sued Illinois because it was dumping sewage into the water supply of St. Louis, about 300 miles downstream, but the case was dismissed when evidence was presented to show that there was no health menace. Conditions were intolerable in the upper Des Plaines River and in the parallel Chicago Sanitary and Ship Canal and the Illinois and Michigan Canal, however, and the City of Chicago had to install sewage-treatment plants through which the waste water had first to pass.

Such debates among communities reflected in part the incomplete understanding of the nature of disease and the technical details of how to design a proper sewage system. Only with the maturation of sanitary engineering and parallel developments in water quality control, including a better understanding of the microbiology and chemistry of drinking water, was more general agreement about effective design criteria and safe levels of contamination established.

DESIGN PROBLEMS

The layout of a water supply system, whether in ancient Rome or in Victorian England or late-nineteenth-century Chicago, has no unique solution, and in this regard it is a problem analogous to laying out any transportation network—roads, canals, or railroad tracks. The start and finish of the route are generally given by the location of the places between which the something (water, carriages, boats, or trains) is to be moved, and it is the job of the engineer to select economical, effective, and acceptable routes among the infinite possibilities between the termini. While scientific and mathematical principles can help translate the problem into analytical terms for conciseness of statement and ease of discussion, it is seldom only a simple matter of finding a solution to an equation, the way the quadratic formula gives the roots of a second order algebraic equation or calculus locates the minimum of a function.

Consider the problem of getting water from a source of supply on one side of a mountain, such as the side where the rain falls heavily, to the reservoir serving a city on the leeward side, which receives relatively little rainfall (see Fig. 8.4). There are numerous ways that the water can be transported from the source, located at A, and the reservoir, located at M. These range from bucket brigades of people to combinations of pumps, pipes, siphons, aqueducts, and tunnels.

For example, a pump might be used to raise the water to a certain level and a tunnel might carry it through the mountain. The tunnel should exit the mountain no lower than the maximum reservoir level desired, which is determined by the geometry of the reservoir and the volume of

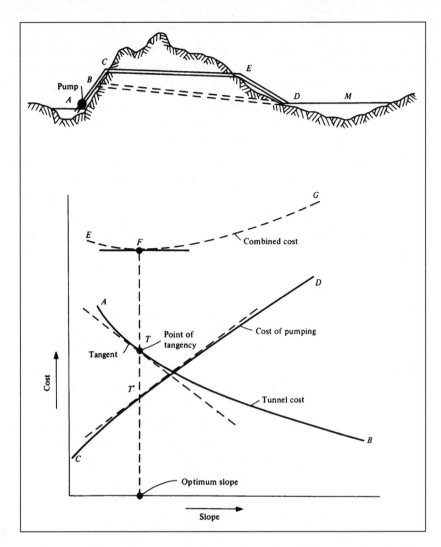

FIGURE 8.4 Schematic diagram defining (top) the elements in a pump/tunnel design problem and (bottom) cost curves for the problem

water that it is designed to hold. The cross-sectional shape and size of the tunnel are related to the desired flow rate, and can be analyzed with the tools of the engineering science of hydraulics. If any kind of solids will be suspended in the water passing through the tunnel, its slope should be sufficiently steep to maintain a velocity that will prevent precipitation of the solids and the disposition of silt, which might in time clog the tunnel. The total length of the tunnel will depend not only on its slope but also on exactly where it cuts through the mountain, which will have different kinds of rock at different locations. The different geological conditions will affect how, and how fast, the tunneling process can proceed, which in turn will affect the cost. All such considerations are summarized in Fig. 8.4 by the curve AB, which is nonlinear because of the complex way in which the various factors contribute to the total cost.

The slope of the tunnel affects where its entrance is located, with steeper slopes requiring that the source water be pumped to higher elevations before gravity will carry it through the tunnel. Larger capacity pumps naturally cost more to purchase, install, and operate. The total cost of pumping for this example is shown as curve CD in Fig. 8.4. By adding the ordinates of the curves representing the cost of tunneling and the cost of pumping, the combined-cost curve EG can be constructed. The optimal pump/tunnel combination, as determined by the criterion of minimal cost, is shown by point F on the combined-cost curve. A perpendicular dropped to the axis of abscissas gives the slope of the optimum solution to the problem.

An alternative way of establishing the minimum cost is to locate the points T and T' on the curves of pumping and tunnel cost, respectively, whose slopes are inversely proportional to each other. Before the advent of digital computers, this kind of calculation was done by trial and error, estimating the slope of the tangent on the graph itself, but now the whole process can be programmed into a digital computer for automatic iteration to any desired degree of accuracy.

MATHEMATICAL AND COMPUTER MODELS

Getting a supply to a reservoir is only one part of the problem of designing a reliable water distribution system. In a real system, one or more reservoirs is interconnected by means of a network of pipes of various diameters and lengths delivering the water to homes, businesses, and other destinations. Before such a system exists, an engineer must establish or estimate the required difference in elevation (known as the "head") between different parts of the network, the location of the pipes and the connections among the pipes, their diameters and lengths, the number and kinds of valves and pumps that might be needed or desired, and the resistance to flow (due to the roughness of the pipes and the presence of various bends and other fixtures) encountered between reservoirs and outlets.

In laying out a new piping network, an inexperienced engineer might follow closely an example of an existing installation that seems to satisfy similar requirements. If there are no appropriate examples, perhaps because of the uniqueness of the new problem, then a combination of existing networks might be studied and the best features of each adopted to the new design. An engineer who has designed networks previously will have gained sufficient experience to be able to estimate rather well the required size of pipes and other components in the network in order to deliver water at its destinations with the appropriate pressure. Whether the engineer is young or experienced, problems with existing networks will need to be especially noted and taken into account in designing the new system. For example, an inadequate volume of flow might indicate the need for larger diameter pipes, and difficulties encountered in cleaning an existing system might call for installing more manholes in the next system.

The evolutionary designs of piping networks could proceed by trial and error over time, but a more analytical strategy is one of the main things that distinguishes modern engineering. An existing or proposed piping network can be approached in much the same way that a physical scientist might approach a system found in nature, namely, by constructing a mathematical model of it based on general principles. In a given piping loop in an ideal network, for example, we know that the mass of water combined or divided at a junction where the ends of several pipes come together (known as a "node") is conserved, that is, the net inflow is equal to the net outflow. A continuity equation can thus be written for that node, and if a piping loop has a certain number of nodes, then that same number of continuity equations can be written for the loop.

Additional equations can be written to express the fact that energy is balanced around a piping loop, taking into account the potential energy associated with the heads of reservoirs, losses through pipes due to friction and change in elevation, and the addition of energy, as through a pump in the loop. Friction losses can be expressed analytically as functions of the pipe characteristics of length, diameter, and roughness, as well as of the rate of flow through the pipe. Such losses are nonlinear functions of the flow rate, so that, for example, doubling the flow rate more than doubles the energy loss in the pipe. These kinds of considerations lead to a system of algebraic equations equal in number to the number of pipes in the loop. Collectively, these equations are said to be a (mathematical) model for the piping loop, and a solution to the equations constitutes a prediction of how water will flow in a real system.

Since the equations modeling a piping loop are nonlinear, they cannot be solved directly or exactly, but various methods have been devised to approximate their solution as accurately as needed for engineering decisions to be made. For any solution to be properly interpreted, however, it is as important for the engineer to understand the assumptions behind, and the limitations of, the models constructed as it is to understand when and how a particular mathematical model can be solved.

Before the advent of digital computers, considerable thought and effort went into developing techniques for the approximate solution of equations such as those representing water flow around a piping loop. Employing these techniques was often a long and tedious process, requiring much repetitive calculation by hand. When extremely large and complex models were employed and large systems of equations had to be solved as accurately and efficiently as possible, the calculations were often done by teams of individuals, each of whom did only a specialized part of the calculation and passed the result on to others on the team who needed that number before proceeding with their own specialized part of the calculation. The individuals in such teams were known as computers, and they gave their name to the electronic machines that have revolutionized the way such calculations are done.

At the heart of the electronic digital computer are still the same fundamental methods of mathematical modeling and approximation that were employed by engineers and teams of human computers that laid out piping systems, electrical transmission networks, railroads, and other parts of our infrastructure that still serve us today. While the computer has enabled more ambitious models to be constructed and solved in a fraction of the time it would have taken in precomputer days, engineers still must understand the principles behind the construction and solution of those models so that their assumptions and limitations are clear. Otherwise, engineers would be unthinkingly operating computers that were for all practical purposes black boxes into which an input was inserted and from which an output was taken. It is to such a situation that the familiar adage "Garbage in, garbage out" is often applied. Without an understanding of the principles of the system being modeled or the nature of the workings within the black box, engineers would be unable to exercise sound judgment about the worth or reliability of the input or the output.

One example of how poorly a system designed by computer can perform was provided by the automated baggage-handling system at the Denver International Airport, a case in which not a continuous flow of water but an irregular flow of discrete pieces of luggage had to be distributed around a network. The airport's opening was delayed for over a year because, in testing the system, bags were being crushed, and the automatic cars they were to be delivered to and from were not connecting properly. The computer model of the system failed to take into account such things as the aerodynamics of the carts and the airflow patterns that flipped lighter luggage up and onto the floor. Among other early problems, there was also an improper matching of car availability and amount of baggage to be moved, a situation analogous to too-small sewers. Though much of the embarrassing performance was later blamed on electro-mechanical hardware problems rather than on the software that operated the system, the fundamental difficulties stemmed from a failure to appreciate that the computer model did not reflect the real system accurately enough. In addition, the system designed for Denver was over ten times as large as any previously built. Such a large scaling up in a single step is something experienced engineers know can be full of surprises, and it is seldom done on critical projects.

Another example is provided by the twin towers of New York City's World Trade Center, which at 110 stories high were the tallest buildings in the world when they were completed in 1973. Among the nonstructural problems faced in designing them was getting an adequate water supply to the top floors. Though the observation deck atop one of the towers provided a new and spectacular bird's-eye view of Manhattan and its environs, including New York Harbor, the Statue of Liberty, and the historic bridges over the East River, some early visitors to the restrooms on the top floor found them to be unsanitary eyesores when the water pressure was inadequate to flush all the waste into the sewer system.

WATER QUALITY

Well into the late nineteenth century, there were two major theories of the causes of diseases like cholera and typhoid fever. One theory, supported by those known as contagionists, held that disease was spread by contact. The other, supported by anticontagionists, held that diseases came from bad air, such as that emanating from soil in which things were decomposing. A lack of clear understanding of causes led to practices that were more harmful than helpful to those struck down with disease.

During the early years of the construction of the Panama Canal, begun by the French in 1880, a large number of workers became ill with malaria and died. The disease was believed to be caused by poisonous marsh gas, as the word itself suggests. ("Malaria" derives from the Italian, *mala aria*, which means "bad air"; the French word for the disease is *paludisme*, which means "marsh fever.") The widely held, centuries-old miasma theory explaining the cause of the disease was supported by the evidence that malaria occurred mainly in hot, humid climates like the Isthmus of Panama, where the abundant growth and decay of vegetable matter gave off gases. Yellow fever, another common disease in the tropics, was believed to be caused by winds blowing across sewage, decaying animal matter, or infected patients.

Suspicions that mosquitoes rather than the air transmitted malaria and yellow fever were put forth as early as the 1850s, but not until the 1880s was a full-blown theory advanced. In the meantime, French hospitals for treating malaria patients during the construction of the Panama Canal were kept fresh with the abundant scent of flowers, whose pots were surrounded by pools of water to keep ants and crawling insects from getting to the flowers and eating them. The legs of hospital beds were also placed in pans of water to keep crawling insects from reaching patients. These practices provided the stagnant water that was the perfect breeding ground for the very mosquitoes that caused the disease. The Americans took over the canal project after the French had abandoned it (in part because of their inability to control the tropical diseases), and standing water of all kind was sprayed with a film of oil to kill mosquito larvae and keep down the mosquito population. By the end of the project, deaths due to the disease were reduced almost to zero.

Among the factors near the end of the nineteenth century that accelerated understanding of the true causes of infectious diseases was the discovery of the existence of microscopic bacteria and the development of the means for isolating and characterizing them by Louis Pasteur in France and by Robert Koch in Germany. Even before the existence of bacteria was firmly established, contaminated drinking water was known to be a cause of disease. In a famous bit of public-health detective work, Dr. John Snow, who was a pioneer in anesthesia, tracked down the source of a cholera outbreak in 1854 by noting that those affected lived in the vicinity of a single source of water in London, while those who worked in a nearby brewery and who got their water from the brewery's own supply remained unaffected. Although both groups were breathing essentially the same air, those who were using the Broad Street Pump were drinking water polluted through the surrounding soil. When the pump handle was removed at Dr. Snow's suggestion, the epidemic subsided.

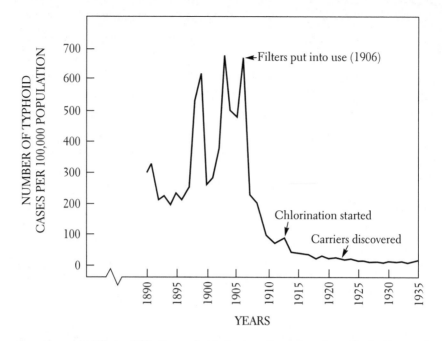

FIGURE 8.5 Effects of filtration and chlorination of water supply on the incidence of typhoid fever in Philadelphia

With increased understanding of the causes of disease and the ill effects of the presence of microorganisms in drinking water supplies, sanitary engineers could design treatment processes and facilities that removed microorganisms from the water before it was introduced into the main piping network of a distribution system. The dramatic effects of filtration and chlorination on the incidence of typhoid fever in Philadelphia is shown in Fig. 8.5. Sanitary and, later, environmental engineers became increasingly interested in longer-term risk factors beyond those presented by microorganisms. These included risk factors associated with trace metals, especially lead, which might be introduced by pipe erosion and corrosion, or chromium and mercury, which might be introduced into water supplies from manufacturing or mining operations.

Continuing studies related to the pollution of water, air, and other components of the environment brought increased attention to public health and ecological issues. The effect on public policy was made official by the passage in 1970 of the U.S. National Environmental Policy Act, and the establishment soon after of the Environmental Protection Agency (EPA). This combined into one independent government body an organization that took over responsibilities formerly distributed among various separate agencies. With the establishment of the EPA, it was easier in theory to coordinate programs that oversaw the control and regulation of matters of environmental pollution in such diverse but often interrelated areas as pesticides, solid waste, air, and water. The subsequent passage of such legislation as the Clean Water Act (first passed in 1972 and revised in 1977), the Safe Drinking Water Act (1974), and the Resource Conservation and Recovery Act (1976) expanded the federal role in pollution control and in research and development in support of that goal. Such concentrated power and centralized control over matters that are still not fully understood naturally have their advantages and disadvantages, their proponents and opponents. Subsequent shifts in political priorities and public policy debates have caused the role of the EPA to change over time.

Just as continuing research has produced evolving theories and ongoing debates within the medical scientific community about the carcinogenic effects of such things as asbestos, so there remained in the late twentieth century some open questions about the exact nature of such global environmental phenomena as the ozone layer and acid rain and how to deal with them, let alone questions about the safe levels of pollutants in water supplies. Engineers must be aware at all times of the status of such issues and standards and their proper interpretation.

OTHER PROBLEMS

Among the legacies of the Industrial Revolution have been numerous new agents and substances that have threatened water supplies and the environment generally. While pollution was not unknown in past centuries, as the water problems of ancient Rome and Victorian London demonstrate, with the late twentieth century came an increasing awareness among engineers, scientists, and citizens generally that pollution problems on a global scale were growing with the world's population. Increasing awareness that human actions could have potentially devastating environmental effects was crystallized by the publication in 1962 of Rachel Carson's *Silent Spring*, which called attention to the toll that the increasing use of chemical pesticides was taking on animal life, particularly birds, and to the potential threat posed to human health as well. *Silent Spring* brought a heightened awareness of the delicate ecological balance in the world and provided an impetus to the environmental movement generally.

Other environmental degradation, though perhaps less immediately noticeable, was also taking place underground, in the groundwater supplies that so many people did or would rely upon. Long abandoned and forgotten, old buried storage tanks containing gasoline, oil, and toxic wastes were beginning to corrode and release their contents into the ground, where the chemicals diffused through the soil and found their way into groundwater streams. Practices once carried out without a second thought, such as using strong chemical solvents to clean airplanes as they stood on the runways of vast military airbases, also had led to enormous amounts of pollutants seeping into ground that had been pristine, if not sacred, in earlier times.

Such past practices, not necessarily done with any malice at the time, have produced new and challenging problems for environmental engineers and hydrologists. Considerable scientific, technical, and computational obstacles must be overcome in developing models of how pollutants move through and threaten groundwater in a vast variety of geological conditions, ranging from strata of sand and clay to masses of fractured and faulted rock. The development of such models must necessarily take place in a context where only limited data are available about the history of a site, the concentrations of pollutants, and the geological map of the media in which the groundwater is flowing. It thus takes clever and sophisticated analytical approaches to predict flow behavior accurately enough to forewarn water users of unacceptably dangerous conditions or to propose possible strategies to preclude such conditions.

The design and implementation of effective remediation strategies that can check or reverse the effects of groundwater pollution are as much engineering problems as are the design and development of a new reservoir or water tower. The end product in a groundwater remediation program is likely to be a system of underground dams or barriers to check further infiltration of pollutants or a system of strategically located wells interconnected with pumping and treating equipment that is designed to remove the pollutants. Perhaps to offset the less tangible end of the design, computer models that lead to such systems are often accompanied by elaborate colorful graphics depicting the motion and concentration of underground plumes of pollutants over long periods of time.

Providing an adequate water supply, disposing of waste water, and dealing with the pollutants in both of these hydraulic systems are among the most challenging problems facing engineers of the twenty-first century. Their analysis and resolution necessarily involve a wide range of knowledge: of the chemical, physical, biological, and geological sciences; of the engineering sciences of fluid mechanics and hydrology; and of mathematics and computer science. In addition, sound engineering judgment must be applied to the design and development of proposals that are reasonable, effective, economical, and of optimal benefit to society, as all engineering approaches should be.

Large engineering projects usually have a long history, which in some cases stretches decades beyond any single engineer's career. There can be innumerable competing proposals and false starts, as well as long periods of inactivity and distraction caused by the need to work on other projects or by economic or political conditions seemingly beyond anyone's control. Elements of luck and timing enter in as well, as they do in any human endeavor. The story of the origins, design, financing, and construction of just about any major bridge can serve as a model for demonstrating this long and tortuous history and the many interrelated complexities that go along with all large engineering projects.

From ancient times to the Industrial Revolution, there prevailed a long and solid tradition of building bridges of stone and timber. Timber bridges were not only common in their own right, but a timber bridge of sorts (known as centering or falsework) had to be erected as a supporting form before a stone bridge could be put in place (see Fig. 9.1). Thus, in the late eighteenth century, when a new bridge was needed in the Severn Valley in England, there was a natural caution against the idea of using iron, despite the fact that the Darby family had made the valley well known for its iron production. The Darbys used the locally abundant coal residue known as coke in place of scarce timber in the smelting process, and the lack of timber made iron for bridge construction even more sensible. Yet the sketches that survive of early proposals for an iron bridge across the Severn River emphasize the strong bias toward traditional materials that had to be overcome.

One of the sketches shows the iron cast into stonelike voussoirs, and the other shows iron mimicking timber. The bridge as built (Fig. 9.2) has the familiar overall appearance of a semicircular Roman stone arch bridge, while the assembly details are reminiscent of contemporary timber bridge construction. Such extra-technological deference to traditional

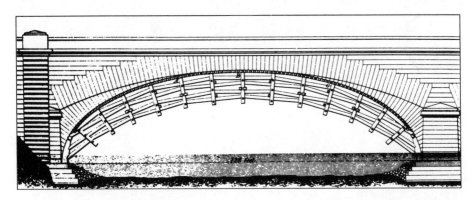

FIGURE 9.1 Falsework in place under a stone arch bridge

materials and methods of construction overcame the resistance to technological change promoted by such vested interests as stonemasons, carpenters, and ferry boat operators. In the final analysis, however, clear and effective communication, ameliorating the sense of threat and uncertainty that the new material prompted, was a crucial factor in getting Iron Bridge built.

Once the bridge was successfully erected, its real advantages could speak for themselves. Because the cast-iron sections were so large and self-supporting, the bridge was put up in such a short time that there was a minimum disruption of river traffic. The promise of iron as an effective bridge material was further emphasized when Iron Bridge was the only bridge across the Severn to survive the flood of 1795. Stone bridges, because of their relatively small opening relative to their massive abutments and silhouettes, tended to clog up and act like dams under flood conditions. Timber bridges, although more open, tended to wash away because of their lack of strength against the force of water and trapped debris. Iron Bridge, on the other hand, being both open and strong, allowed the raging flood waters to pass through its lacy structure, which was seen eventually to possess an aesthetic of its own.

With the triumph of Iron Bridge, bridge designers increasingly began to think of iron as an innovative new material with which to work. But, whereas the design of Iron Bridge, based on the long-tested arch principle, required little sophisticated analysis of the forces acting on it, the bolder designs for iron bridges that were proposed in the early nineteenth century had little precedent upon which to argue for their safety. Suspension bridges, for example, used wrought iron in tension instead of cast

FIGURE 9.2 Iron Bridge, completed in 1779, across the Severn River

iron in compression, and the structural action involved had no counterpart in stone or timber construction. Engineers proposing suspension bridges had to rely in public on their drawing, writing, and speaking talents as much as they did in private on their analytical and political abilities. Whereas they may have addressed to their satisfaction any strictly technical questions at the drawingboard or through experiments, the questions of strength, safety, economics, aesthetics, and the like that were raised by citizens, investors, politicians, and other lay persons required a talent for persuasive speaking and writing.

Proposals for a variety of different bridge types, and reports of progress on their construction, were models of nineteenth-century engineering. They were made possible by steady progress throughout the century in understanding beam action and in analyzing structures, and this new understanding was often conveyed to the lay public. However, in the 1840s in Britain, suspension bridges were not considered viable options for railroad structures, despite the great demand for crossings as rail routes were expanded. This poor stature of suspension bridges was attributable in part to the bridges' lack of stiffness and susceptibility to collapse in high

As a young engineer, Othmar Ammann worked for Gustav Lindenthal, whose design for a massive suspension bridge between New York and New Jersey across the Hudson River was so ambitious and expensive that in over thirty years it had not gotten sufficient backing to progress very far. After trying unsuccessfully to get Lindenthal to redesign his bridge to less monumental proportions, in the early 1920s Ammann began to work independently on a more modest proposal for a Hudson River bridge.

Without informing Lindenthal, Ammann communicated his ideas to the future governor of New Jersey, who then promoted it as part of his administration. When Lindenthal found this out through newspaper accounts, he accused the younger engineer of unethical behavior. Did Othmar Ammann act in an unethical way?

winds, which were known to engineers and laypersons alike. As a result, many alternative bridge designs were developed in Britain, including the iron tubular span exemplified in the Britannia Bridge. The extreme weight and cost of the Britannia, however, led to the evolution of open truss girder designs, which were to be taken to extremes of lightness in the Tay Bridge, as we shall see.

Railroad bridge building in America followed a different evolutionary route, in part because the persuasive qualities of John Roebling's personality and writing complemented his engineering talents and acumen. Roebling's Niagara Gorge Suspension Bridge, completed in 1854, provided an incontrovertible counterexample to the British hypothesis that a suspension bridge could not carry railroad traffic. He went on to conceive and propose the landmark Brooklyn Bridge, and the story of the many long years that Roebling spent trying to convince others of his plan, after he had first persuaded himself, is one of the great dramas in the history of engineering. It often serves as a paradigm for all engineering projects precisely because the vicissitudes of that great project are far from unique.

But we might just as easily take as paradigmatic the story of Othmar Ammann and the George Washington Bridge, which was prefigured for almost a century by numerous plans for crossing the Hudson River at New York City. Ammann's pragmatic flexibility enabled him to achieve what his mentor, Gustav Lindenthal, could not, in large part because he could not adapt his railroad-age design for a behemoth of a bridge to the changing needs of a society that was developing a growing love for the automobile. In the early 1920s Ammann and Lindenthal parted ways, and the younger engineer was able to persuade political and business groups on both sides of the Hudson that his light and graceful $30 million bridge between Fort Lee in New Jersey and 179th Street in Manhattan was the route to go, even though technically it represented effectively a doubling of the longest suspension bridge span then in existence. Both Lindenthal and Ammann have left a voluminous legacy of written documents arguing the case for bridges, both built and unbuilt.

SAN FRANCISCO BRIDGES

Among the major bridges of the world whose story is not as well known as it should be is the crossing of the bay between San Francisco and Oakland, California. Officially called the San Francisco—Oakland Bay Bridge, this enormous structure of over eight miles in total length has been overshadowed almost from the beginning of its operation by its neighbor, the Golden Gate Bridge, just a few miles to the northwest. Ideas for bridges at each of these locations go back to the nineteenth century, including a curious proposal issued in the form of a proclamation in 1869 by a fellow who declared himself Norton the First, Emperor of the United States and Protector of Mexico. Joshua A. Norton is believed to have been a fortuneseeker who went to California during the Gold Rush of 1849, struck it rich, but then lost everything, including his mind. He disappeared for a while but then reappeared as the self-declared emperor whose seat of power was in Oakland, where local people humored him and honored the paper money he issued. The fact that he proclaimed that a bridge be built from Oakland to Sausalito via San Francisco suggests that the idea was current in his time. However, it was not until the early decades of the twentieth century, when major bridges were being constructed all across the country, that proposals for various bridges in the San Francisco Bay area began to receive more serious attention.

Engineers who found themselves in California on other business became familiar with the need for bridges across the bay and its outlet to the sea—the strait known as the Golden Gate. One of these engineers was Joseph Strauss, who was president of his own Chicago-based Strauss Bascule Bridge Company. A bascule bridge is one whose roadway rotates on a hinge so that the span can be lifted out of the way of ships that need to pass. In this way the bridge can be built with a low clearance across a river, and thereby not require a lot of land on either side for long and high approaches. Strauss had become successful by designing and building such movable bridges in built-up and crowded cities like Chicago, where land for bridge approaches was very expensive, if available at all. With his experience in patenting and building moveable bridges, Strauss was also able to design an amusement ride known as the Aeroscope, in which passengers were carried up in a car as large as a two-story house and revolved at the end of a 265-foot cranelike boom over the grounds of the Panama-Pacific International Exposition, which was held in San Francisco in 1915 to commemorate the opening of the Panama Canal.

In that same year, the Strauss Bascule Bridge Company was awarded a contract to replace an old swing bridge across a San Francisco creek with a new bascule bridge. The Fourth Street Bridge, as it is known, is generally considered an ugly and purely functional design, as were many such bridges, with a heavy concrete counterweight suspended overhead. In any event, business travels relating to the Aeroscope and the Fourth Street Bridge caused Strauss to cross paths with the San Francisco City

Engineer, Michael O'Shaughnessy, who dreamed of having a bridge across the Golden Gate.

Within a few years, Strauss had produced an uninspired design for a Golden Gate bridge. However, no matter how unattractive it was, Strauss's cost estimate of $17 million and his projection that toll revenues would in time more than cover that entire cost made the proposal very attractive to proponents of a bridge. Furthermore, Strauss's political savvy and entrepreneurial spirit, which drove him to talk up his proposal before any local civic or political group that would listen, eventually led to the dream becoming reality. The human story of Strauss and the Golden Gate Bridge is a long and acrimonious one, in which Strauss and O'Shaughnessy eventually became adversaries and in which Strauss dismissed and denied credit to his assistant engineer for design, Charles Ellis. However, in spite of the bridge's unglamorous beginnings and the difficult interpersonal relationships that ensued during its design and construction, the Golden Gate Bridge stands today as one of the most admired bridges in the world, at least in part because of its dramatic setting and the strikingly simple and beautiful design that eventually evolved.

The San Francisco–Oakland Bay Bridge, on the other hand, has a less unified appearance and occupies a site that does not lend itself to dramatic sunsets or to views that encompass the entire bridge. In contrast to the Golden Gate Bridge, whose single dominant 4200-foot main span was at one time the world's longest, the Bay Bridge is really made up of two distinctly different kinds of spans, each of which constituted a major bridge project in its own right, plus a significant tunnel through an island that rises high out of the water in the middle of the bay, a couple of miles from either shore. Although the construction of the Bay Bridge across deep water that was heavily traveled by ships presented enormous technical challenges to its engineers, it is not as widely recognized for the achievement it is because of the social and aesthetic accidents of its construction and appearance, and because of its being overshadowed by the contemporary project of the Golden Gate Bridge.

EARLY PROPOSALS FOR A BAY BRIDGE

With most large bridge projects, a variety of proposals are put forth before a final design is settled upon. Just as Strauss's early and ungainly design for the Golden Gate Bridge was fortunately modified over the course of its design time to become an object of beauty as well as a technical achievement, so the Bay Bridge evolved slowly from early proposals to its final form. Among the more serious early proposals was one put forth by the civil engineer Charles Evan Fowler. Fowler was a Seattle-based consulting engineer who, like many of his contemporaries, dreamed of building the largest bridge in the world. Just as Strauss had become acquainted with the problem of bridging the Golden Gate while doing other business in San Francisco, so Fowler became interested in the problem of a bay bridge when he was engaged in designing the steelwork for one of the city's earliest steel-frame buildings. Like many a consulting engineer, Fowler was always looking for opportunities to secure new and prominent commissions. He prepared and printed at his own expense a booklet that laid out the problem of bridging the bay and presented his design for a solution.

Fowler published his booklet, entitled *The San Francisco–Oakland Cantilever Bridge*, in 1914. At that time, tunnel schemes for crossing the bay had been estimated to cost at least $25 million, and it was necessary to explain why any bridge costing a considerable amount more was a better choice. In fact, because of technical surprises that seemed invariably to accompany tunnel projects and because of the limited traffic capacity their necessarily small diameter tubes provided, comparative studies of tunnels and bridges were a matter of some heated debate not only in San Francisco but around the world. Thus, although Fowler's bridge proposal carried a price tag of $75 million, he argued that it was in the final analysis cost-effective and proceeded to describe it in the context of known bridge projects of the time.

Among the big decisions that faced bridge engineers in the early part of the twentieth century was the basic structural form that they would employ for long spans. Such a span would be required to bridge the deep-water distance between the city of San Francisco and the large island two miles out in the bay known variously as Yerba Buena Island or Goat Island or, more recently, as Treasure Island. This last name, strictly speaking, refers only to an artificial island formed adjacent to the large rock outcrop that comprises the natural island. Indeed, Treasure Island was created as the site of the 1939 West Coast World's Fair held to commemorate the engineering achievements of both the Golden Gate Bridge and the San Francisco–Oakland Bay Bridge.

Because the water was approximately 200 feet deep, the construction of foundations and piers to serve as bridge supports for a crossing would be costly and dangerous. It was thus desirable to have as few supports in the water as possible, and this meant that the superstructure should have spans as long as possible. In 1914, when Fowler was writing, the suspension bridge with the longest span was the 1600-foot Williamsburg Bridge in New York City. Although as early as the late 1880s engineers like Lindenthal had proposed bridges with suspended spans of the order of 3000 feet, such extra-long-span designs did carry very large cost estimates. Furthermore, they were so far beyond the experience of bridge building projects up to that time that the wisdom of making such large leaps in the state of the art was seriously questioned.

CANTILEVER BRIDGES

Fowler wrote in his report that the problem of bridging the bay between San Francisco and Oakland had been "uppermost" in his mind for a quarter of a century. That would have put his first thoughts on the matter in the late 1880s, when cantilever bridges were attracting a lot of attention among engineers. A cantilever bridge is a self-supporting structure that can be built out as large individual cantilever beams from piers or towers, thereby requiring no scaffolding underneath (see Fig. 9.3). This is an attractive design and construction technique because it eliminates the need to erect scaffolding that might obstruct shipping during bridge construction and that would certainly add to the bridge's cost.

The idea of a self-balancing cantilever was novel in the late 1880s, however, and a very large one was being built in Scotland only because other types of bridges had fallen into disrepute after a famous bridge collapse. The need for long bridges on the east coast of Scotland had developed when the North British Railway wished to eliminate the ferrying of railroad cars across the wide estuaries known as the Firth of Forth and the Firth of Tay. The bridge across the Tay was completed first, in 1878. It was a long, sinuous structure with many piers in the relatively shallow water. The most striking feature of the bridge was the high girders that rose, in almost 250-foot spans, above the main shipping channel. One night late in 1879, these fell or were blown into the water during a violent storm. About 75 people were killed in the horrendous accident, and there was a major inquiry into why it had occurred.

A Court of Inquiry found that the bridge's engineer, Sir Thomas Bouch (he had been knighted for the project), had been negligent in his design and in his supervision of construction. The force of the wind on the light bridge had been grossly underestimated by Bouch, and one theory of the failure was that the high girders and the train crossing them were simply toppled off their supports, to which the girders had not been sufficiently

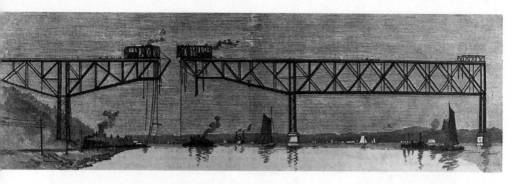

FIGURE 9.3 The railroad bridge at Poughkeepsie, New York, under construction in the late 1880s, illustrating the cantilever principle

anchored. Another finding was that the iron columns on which the bridge stood were made of inferior castings, full of holes that weakened the structure. As with many early structural failures that left a mass of twisted metal, the exact cause remained somewhat debatable, but Bouch's reputation was irreparably damaged. A replacement bridge was designed and built across the Tay, but the task was to be given to other engineers.

At the time of the Tay Bridge failure, Bouch had also been overseeing the construction of a more ambitious bridge across the Firth of Forth, about 50 miles to the south. His design for that crossing had been a long-span suspension bridge, because the water was deep and he wanted to put as few piers in the water as possible, in order to work within allowable depths and to keep costs down. Not surprisingly, when the Tay Bridge fell, construction on Bouch's Forth bridge was halted. After he was found responsible by the Tay inquiry, Bouch's design was abandoned altogether. The task of designing a new bridge for the Firth of Forth was eventually given to the well-known and experienced British engineer Sir John Fowler. Since a major concern of the railroad company was to win back among passengers some degree of confidence in its bridges, a suspension bridge design, which was not generally favored in Britain anyway, became even more undesirable because it had been favored by Bouch. Sir John and his young assistant engineer, Benjamin Baker, thus came up with a relatively new idea for a British bridge—a cantilever.

The errors made in the design and construction of the Tay Bridge were described in considerable detail in the report of the Court of Inquiry, and the Tay accident led to the development of the massive cantilever bridge across the Firth of Forth. This bridge was the subject of many public lectures by Baker, and of numerous written reports reproduced around

FIGURE 9.4 Anthropomorphic model of the cantilever bridge over the Firth of Forth

the world. Because the Forth Bridge was so ambitious a design, especially in the wake of the collapse of the Tay, which was on the same railway line, it was necessary for Baker to communicate to the public the principles of cantilever construction.

Like most ideas, the cantilever bridge was not entirely new. Such bridges had been built in the Far East for some time, and some had of late been constructed in Germany and America (see Fig. 9.3). Thomas Bouch himself had even constructed a lesser-known one in Newcastle. However, the cantilever that Fowler and Baker proposed for the Firth of Forth was to be enormous, spanning about twice the distance of any then in existence. Because they were doing something so daring, Baker took great pains to explain the principle of the cantilever in the numerous public lectures he gave while the bridge was under construction. To illustrate the structural ideas behind the design, Baker employed a human model that consisted of two people on chairs and supporting a third person on a seat suspended between them (see Fig. 9.4). The arms of the people on the chairs and the struts that they held represented the cantilevers,

with the counterweights of bricks providing balance. The swing represented a suspended portion of the bridge, but it looked nothing like the
kind of suspension bridge Bouch had proposed, nor did it behave in
principle like it. The distance from chair to chair thus came to be referred
to, incorrectly, as the span of the cantilever itself.

The Firth of Forth Bridge was immensely successful. It became famous
throughout the world and had many imitators, and by the end of the
nineteenth century a cantilever was the bridge of choice for many designers, builders, and purchasers of bridges. When it became desirable to
bridge the St. Lawrence River at Quebec, a cantilever with spans even
longer than the record 1710 feet of the Firth of Forth was designed. Construction of the south arm of the Quebec Bridge had reached a distance
of about 600 feet over the river in August 1907 when the bridge suddenly
collapsed, killing about 80 workmen. Though the Quebec bridge was
found to be grossly underdesigned, the failure called cantilevers of all
kinds into question. The Quebec Bridge was eventually redesigned as a
stronger structure and was finally completed in 1917. At the end of the
twentieth century its span of 1800 feet remains the record for a cantilever
bridge.

Such troubles with cantilever bridges were not thought to be likely
when Charles Evan Fowler first began to think about bridging San Francisco Bay, for the Firth of Forth Bridge was thought to be a model for
design. Perhaps because of its association with Sir John Fowler, whose
surname he shared, or perhaps because later on Charles Evan Fowler
knew that the Quebec incident was a case of bad design that need not
condemn the whole genre of cantilever bridges, for decades he clung to
the design he had come up with in his youth for a bridge between San
Francisco and Oakland. Fowler proposed a Forth-like cantilever bridge
(Fig. 9.5) with three main spans of 2000 feet each (two 650-foot cantilever
arms supporting a suspended span of 700 feet between them), thus proposing a bridge longer than any in the world. Fowler dismissed still longer
cantilevers of 2300-foot span and a suspension bridge alternative because
of cost and "seismic conditions" in the bay area.

His report included a diagram showing the relative proportions of his
2000-foot cantilever span and a suitably proportioned outline of the
1710-foot spans of the Firth of Forth Bridge, which itself had once been
compared to more familiar structures, such as the Eiffel Tower (Fig. 9.6).
Fowler also pointed out that the profile of his spans was chosen "not only

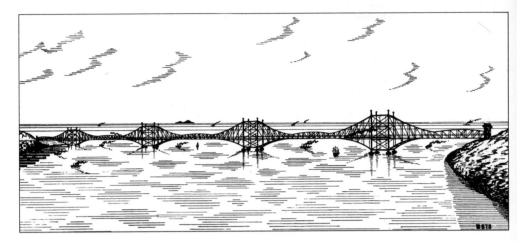

FIGURE 9.5 Charles Evan Fowler's design for a cantilever bridge between San Francisco and Goat Island

for economical reasons, but to provide a structure of pleasing design." Such aesthetic considerations are often very high in the consciousness of engineers designing large, monumental structures.

Although Fowler did not mention the Quebec Bridge in his report, he knew that his readers would recall its relevance. It was well known that the Quebec Bridge failed because its lower chord members, corresponding to the struts held by the seated figures in Baker's model, were simply not strong enough to resist the great compressive forces being imposed upon them. The situation was analogous to those people holding not thick objects like baseball bats but slender ones like yardsticks. Pushing down too hard on such slender sticks, as we can readily verify by a chair-seat or desk-top experiment with a plastic yardstick or ruler, causes a phenomenon known as buckling, in which the stick suddenly springs out from under the pressure. Since such a situation had effectively occurred at the Quebec Bridge construction site in 1907, Fowler wanted to assure his readers that the same thing could not happen in his design. He did this by comparing the area of the lower chords of several well-known contemporary bridges. The Queensboro cantilever bridge, with a maximum span of almost 1200 feet, had recently been completed in New York City with a lower chord area of 1120 square inches. Also in New York at the time, a massive steel arch railroad bridge was under construction across a treacherous stretch of water known as Hell Gate, and its lower chords had an area of 1437 square inches. Fowler's bay bridge was to have chords of

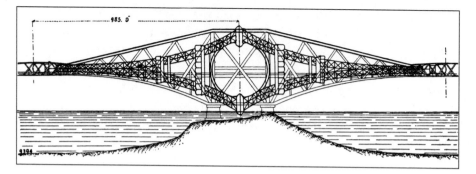

FIGURE 9.6 Scotland's Firth of Forth Bridge compared, to scale, with the contemporary Eiffel Tower

octagonal cross section with an area of at least 3600 square inches, and he assured his readers that areas as great as 6000 square inches could be fabricated.

The great cantilevers proposed by Fowler were necessary to cross the bay between San Francisco's Telegraph Hill and Goat Island, the part of the waterway that required the widest and highest clearance for shipping. Fowler's plan for crossing the island was to excavate to the degree needed to provide a level path for four railroad tracks and two roadways. The considerable amount of excavated earth was to be used to fill in low areas and extend the island into some shallow water to the west, thus reducing somewhat the total length of bridge there. Over the East Bay between Goat Island and Oakland, Fowler proposed using earth fill to extend the land about two miles out into the bay. The remaining two miles of bridge needed to cross the water would consist of about a mile and a half of viaduct, which would essentially be a bridge of many modest spans on piers in relatively shallow water, plus another major cantilever bridge, but with a comparatively modest maximum span.

Proposing a bridge of the kind Fowler did would be only an exercise in structural design if he could not show it to be practical and economically feasible. Matters of practicality included getting railroad trains, vehicular traffic, and people on and off the bridge where it touched land. This was relatively easy on the Oakland side, where there could be a long bridge approach over filled land, thus providing a gradual grade. On the San Francisco side, however, it was another matter to bring traffic down from a bridge that had to be 165 feet above the water to allow ships to pass beneath it. Since the land that might be used for approaches in the city was developed and expensive to purchase and clear, it was desirable to minimize the length of approaches. To reduce the approach distance of railroad tracks while maintaining a reasonable grade, Fowler proposed that the railroad terminals be elevated. He also proposed that large elevators be installed capable of raising and lowering great numbers of pedestrians at a time from a bridge terminal on Telegraph Hill.

The economic justification for the elaborate bridge scheme Fowler proposed lay in the great increase in ferry traffic across the bay that had developed over the decades. During 1873, according to Fowler, there were about 2.5 million ferry passengers. By 1877 the number had doubled to over 5 million. By 1912 the number had reached 40 million annually and was projected to increase at the rate of about 2 million per year for the next ten years, the time estimated to build a bridge. Thus, by the time the bridge could be completed, in the mid-1920s, it would have to have a potential for carrying 60 million people annually.

Of Fowler's estimate that $75 million would be needed to build a bridge and terminals and pay interest on the money borrowed for the construction, two-thirds of the total, or almost $50 million, would be required for the massive cantilevers alone. Still, in Fowler's judgment, the "probable earnings" at the time of the bridge's completion would make it a paying proposition. Not everyone agreed, however, and in his massive two-volume 1916 treatise, *Bridge Engineering*, the prestigious and influential engineer J. A. L. Waddell doubted whether wagon and automobile traffic would pay for the upkeep of the roadway, let alone pay for the interest on the additional cost of construction for it.

To establish his own credentials for such a bold proposal, Fowler had appended to his report a list of his published works and photographs of some bridges for which he had served as chief engineer of design or as a consulting engineer. These included a steel arched cantilever in Knoxville, Tennessee, the towers of the Williamsburg Bridge in New York City, and a triangular railway arch bridge at White Pass, Alaska. Though several aspects of his main bridge design were remarkably similar in type and location to many others that would be presented over the next decade or so, Fowler was not to be involved in the bridge that was eventually built. In part, his timing was unfortunate, in that a second accident during the final stages of construction of the Quebec Bridge occurred in 1916, when the suspended span fell into the water as it was being hoisted into place. This incident cast further doubts on Fowler's judgment in proposing a cantilever bridge whose main spans would be even longer than those of the Quebec Bridge and thus even more difficult to erect. Furthermore, developments in international affairs turned people's minds and society's resources to other matters.

FURTHER PROPOSALS

World War I caused virtually all large bridge projects to be put aside, but in the 1920s schemes for crossing San Francisco Bay were revived. Especially popular were plans capable of relieving the growing traffic problems created by automobiles and trucks while at the same time turning a profit for investors who would build and operate toll bridges. Of course, ferries still carried vehicles across the bay from San Francisco to Oakland and Berkeley to the east and to Marin County to the north, but this mode of transportation was slow and frustrating, and during rush hours or over weekends the lines of cars waiting for a ferry could seem to be interminable. The ferry operators understandably opposed the building of bridges, but bridges were becoming increasing attractive alternatives.

In cities like New York, where colder winters could cause the Hudson River to ice over, thus interrupting ferry service altogether, cries for bridges and tunnels were loud and clear. Thus, the first vehicular tunnel under the river was begun in the early 1920s, amid considerable controversy regarding the nature of its design and construction and questions as to whether poisonous exhaust fumes could be effectively vented. The tunnel came to be named after its chief engineer, Clifford Holland, who died shortly before the tubes connecting New York and New Jersey were dug through, and the tunnel was an immediate success when it was opened to traffic in 1927. It was also loaded to capacity, for by that time the number of automobiles and trucks had grown to such numbers that new tunnels and bridges across the Hudson were already being designed, including the Lincoln Tunnel and the record-setting 3500-foot suspension bridge to be known as the George Washington Bridge, which would be completed in 1931. The successful financing of the Hudson River crossings by the Port of New York Authority, as well as that of other contemporary projects, by means of bonds that would be entirely repaid by the tolls charged to users of a tunnel or bridge, eventually helped gain final approval for projects like those in San Francisco.

Unlike Charles Evan Fowler's report, which proposed an expensive bridge but did not make a sufficiently convincing argument that it would pay for itself, Michael O'Shaughnessy and Joseph Strauss's 1921 proposal for a Golden Gate Bridge was a model of salesmanship. Not only was it better written than Fowler's bay bridge proposal, but it spelled out concisely and explicitly the costs and benefits in terms readily understood by anyone. Though O'Shaughnessy could elicit few formal proposals other than Strauss's for a crossing of the Golden Gate into the relatively unpopulated northern counties, proposals for crossing the bay to the east, where most commuters to the city lived, were numerous. In 1921, for example, thirteen applications were on record seeking franchises from the city of San Francisco to build a bridge to Oakland or other East Bay communities.

However, since San Francisco Bay was such a strategic waterway, the War Department, whose approval would eventually have to be won, held hearings on the matter of a bridge and placed strict limitations on what kind of structure could be built. Among these restrictions were that no bridge of any kind would be approved north of a location known as Hunter's Point, which was south of San Francisco; that no low bridge would be approved north of San Mateo, which was ten miles further south still; and that a 3000-foot-wide open channel had to be left on the San Francisco side of the bay.

The constraints imposed by the War Department, as well as the fact that the cost of a bridge across San Francisco Bay was expected to exceed that of any bridge built to that date in America, kept an easy decision from being made. By 1926, in addition to the cantilever design submitted by Fowler, seventeen proposals were before the San Francisco Board of Supervisors, who were being petitioned for a franchise to build a bridge with private funds. The designs ranged from combination tunnel and bridge schemes to major suspension bridges, and many of the proposals had associated with them the names of some of the most famous engineers of the time. These included J. Vipond Davies, who had been responsible for tunnels under the Hudson River; Ralph Modjeski, who was chief engineer for the suspension bridge then being completed across the Delaware River at Philadelphia, as well as many other bridges; George Washington Goethals, who had become famous as the engineer who finally completed the Panama Canal and who was advocating a tunnel scheme in New York; J. A. L. Waddell, the author of *Bridge Engineering*, who had built, among many others, a famous lift bridge in Chicago; and Gustav Lindenthal, who for almost 40 years had promoted a gigantic suspension bridge to carry rail and road traffic across the Hudson River and who had designed important bridges in Pittsburgh and New York City. Even Joseph Strauss, who at the time was promoting his Golden Gate project, had designed a Sunshine Transbay Boulevard Bridge that was to consist of cantilever spans and a bascule crossing between Hunter's Point and Alemeda. Furthermore, the longer a bay bridge remained unbuilt, the more designs were offered, and in 1928 the number of proposals put forth had reached 38.

SELECTING A SITE

While there was some question as to the legality of San Francisco unilaterally selecting a site and having plans prepared for a toll bridge to be built by a private company, the city's Board of Supervisors proceeded to ask the presidents of four local universities to provide a list of qualified and disinterested bridge engineers to serve in an advisory capacity. Three engineers were to be selected from the list to make "a study and investigation of the proper location of termini, foundations, clearance above the waters of the bay, space between piers, the loads which said bridge should carry, the facilities for traffic at both termini of said bridge, and the probable conditions of traffic incident to said bridge, and the financial problems involved." After so many years of indecision on the part of the city, the presidents of Berkeley, Stanford, the University of Santa Clara, and St. Mary's College were asked to produce a list of names within ten days. Since so many of the most famous bridge engineers in the country were already involved with the proposals that had been put forth, the list submitted contained names that might not have been immediately recognized. However, a distinguished board was appointed, made up of Robert Ridgeway from New York, Arthur N. Talbot from the University of Illinois, and John D. Galloway from San Francisco.

The board of engineers was given only 30 days to deliberate, and they came to the following conclusions in a report filed on May 5, 1927:

1. The most suitable location is from Rincon Hill to Alameda Mole [that is, breakwater], with a second choice from Potrero Hill to Alameda Mole, and a third from Telegraph Hill to Yerba Buena Island, thence to the Key Route Mole [as shown in Fig. 9.7].
2. The longer spans should be two 1,250-ft. cantilevers.
3. Maximum grades should be 6 per cent for vehicles and 4 per cent for inter-urban [railway cars].

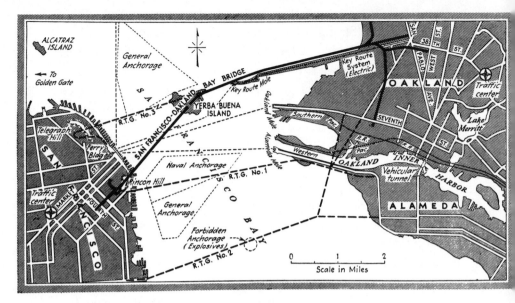

FIGURE 9.7 Map of San Francisco Bay, showing possible bridge locations

4. A double-deck bridge with a 42-ft. roadway on the upper deck and three inter-urban tracks on the lower deck will provide the desired capacity.

The board of engineers recognized a reliable cost estimate could not be made before information was available about the nature of potential foundation conditions beneath the bay. For all of the applications for franchises and preliminary design proposals that had been submitted, there had been remarkably little exploration of where exactly, and how deep, the bridge piers would be located. This should not be surprising, for the proposals were prepared on speculation, and few engineers or financial backers would have wanted to risk the time and money to explore the bed of the bay in any degree of detail before they had some kind of commitment that their efforts would have a reasonable chance of being rewarded. Thus, the board of engineers recommended that the city desiring the bridge explore the preferred site to determine exactly what kind of foundations could be supported and prepare a more detailed design on which to base cost estimates.

In the meantime, the War Department continued to be an obstacle. City officials attempted to deal with it by going to Washington to try to persuade army and navy authorities to relax their restrictions, and when this was not effective, the city got its elected representatives in Washington to introduce legislation to circumvent the War Department. This was not successful either, and further political efforts led to the idea of proposing the bridge as a publicly-owned facility, which would enable it to be financed by revenue bonds in a manner not unlike the way the interstate Port of New York Authority was financing the George Washington Bridge. The idea of public ownership led to the notion of a governmental commission to oversee the project, and in 1929 a representative of Governor C. C. Young enlisted the support of President Herbert Hoover, a graduate of Stanford and a successful mining engineer before he became involved in public service and politics. He soon announced the appointment of a San Francisco Bay Bridge Commission, which came to be known as the Hoover-Young Commission. Legislation also created the California Toll Bridge Authority and designated the Department of Public Works as its agent to design, construct, and operate toll bridges. It further provided for their financing through revenue bonds and appropriated money for preliminary investigations.

The commission put California's state highway engineer, Charles H. Purcell, in charge of the project. Purcell, who was born in 1883 and who attended Stanford and the University of Nebraska, began his engineering career as a bridge designer for the Oregon Highway Department and worked for the U.S. Bureau of Public Roads before coming to California. State Bridge Engineer Charles E. Andrew was given immediate charge of the study. A systematic investigation of possible bridge locations was begun, and borings were made to determine foundation conditions. Detailed traffic studies were also initiated, including compilations of ferry records, in order to have accurate and up-to-date data on which to base projected bridge use and toll revenues. These studies led to the clear conclusion that the best route for the bridge was from San Francisco's Rincon Hill to Goat Island, along the route of an underwater rock ridge that provided a base for the least deep foundations, and thence on to Oakland (see Fig. 9.7), even though that community had been somewhat reluctant to have a bridge interjected into its harbor.

The Hoover-Young Commission received a report of the study, and by the middle of 1930 was able to conclude that bay bottom conditions made it clear that the route over Goat Island was the only technically and economically feasible one and that there should be no more than four main spans over the West Bay and that the two center spans should be high enough to provide a clearance of 214 feet over mean high water.

This part of the bridge was as technically ambitious a construction project as any then under way or contemplated, and yet the final design was to be "such that it will conform with the scenic beauty of San Francisco Bay." The responsibility for designing and constructing such a bridge lay with the California Toll Bridge Authority.

THE FINAL DESIGN

With $650,000 appropriated in 1931, serious design work could begin, which meant that an engineering organization had to be established to look at the various options and details and come up with final plans so that War Department approval could be obtained and construction could begin. Such work necessarily involves a great number of engineers, with different ones focusing on different aspects of the project, which in this case included the West and East Bay crossings, the Goat Island crossing, and approaches in San Francisco and Oakland. Purcell was made chief engineer of the entire project, Andrew was appointed bridge engineer and as such headed the design team, and Glenn B. Woodruff was made engineer of design. As with all good engineering work, an independent group of engineers was established to advise on and check all engineering assumptions and calculations. This board of consulting engineers included Ralph Modjeski, whose Delaware River Bridge had been the longest when completed in 1926; the firm of Moran & Proctor, whose principals Daniel Moran and Carlton S. Proctor had been responsible for the foundations and piers of some of the country's most significant bridges; Leon Moisseiff, who would also work on the Golden Gate and Tacoma Narrows bridges; Charles Derleth, Jr., who was dean of engineering at the University of California in Berkeley; and Henry J. Brunnier, an established San Francisco consulting engineer who had been involved with many important West Coast structures. Consulting architects were also appointed to give advice on the physical appearance of the bridge and its approaches.

The final design of a great bridge is never a simple matter, for there are many different factors to be considered and many conflicting objectives to be met. The result is many possible designs (see Fig. 9.8), and it is engineering experience and judgment that play a key role in narrowing down the possibilities. The first design that was focused upon for the purposes of gaining War Department approval was a four-span symmetrical cantilever arrangement, in spite of the form's history. Because the 1700-foot spans were comparable to those of the well-considered Firth of

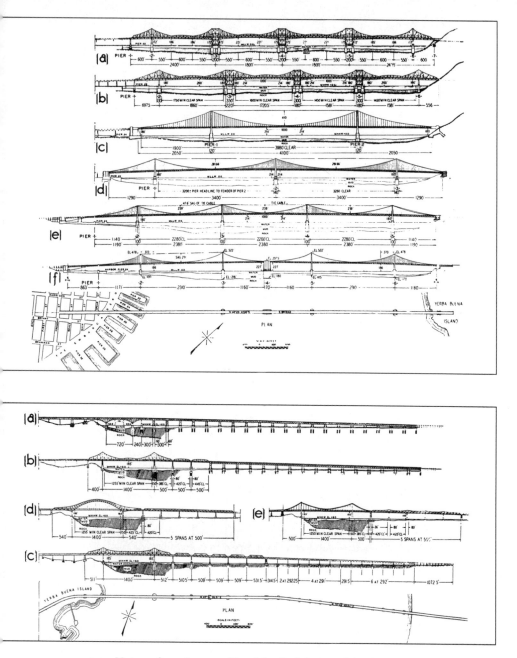

FIGURE 9.8 Various alternatives considered for final design of the Bay Bridge: (top) West Bay crossing, and (bottom) East Bay crossing

Forth Bridge, the feasibility of the design and the clearances for shipping that it could provide were virtually unquestionable, and thus the critical approval was finally obtained. Any subsequent design changes that would be found desirable for technical or economic reasons would then be less likely to have approval withheld.

In fact, even before approval had been obtained for the cantilever design, it became clear to the engineers that a suspension bridge would be more economical, safer to erect, and more aesthetically pleasing. But exactly which suspension bridge design to adopt was still a matter of some thought and judgment, and required further calculations and tests.

It is always tempting for engineers to build a bridge with as long a span as possible, not only to have a sense of grandeur but also to reduce the number of obstacles in the water. Thus, a design with a single suspended span of 4100 feet between two towers would have been in keeping with the growth of suspension bridges while at the same time providing a structure of monumental proportions. In the end it was rejected, however, since it was found to be more expensive and less suited to the location than multiple span bridges. However, since the design of a multiple-span suspension bridge was indeed a departure from experience, special model tests were conducted by engineering professors R. E. Davis at the University of California and G. E. Beggs at Princeton University, in order to guide and check theoretical calculations about the dynamic behavior of such structures. In the end, these tests helped settle on a design comprising two complete suspension bridges in tandem, sharing a common central anchorage. The main spans of 2310 feet still made each of these individual bridges larger than all but the George Washington Bridge recently completed between New York and New Jersey and the Golden Gate Bridge under construction across the bay, but the spans were generally considered otherwise to be relatively conventional in design.

The crossing of Goat Island, on which there were military facilities, was to be through a tunnel of larger bore than any then in existence, making the overall project still more remarkable. As for the East Bay crossing, the wishes of Oakland had to be accommodated so that future development of its port facilities were not jeopardized by a low bridge. The original design calling for a main section comprising a 700-foot cantilever with a 161-foot clearance was modified to provide a 1400-foot cantilever section with a vertical clearance of 185 feet (see Fig. 9.9). Thus the project was to include a major cantilever span that would be the third

FIGURE 9.9 Cantilever span under construction over East Bay

largest in the world, ranking only behind the Firth of Forth and Quebec bridges.

The total cost of $75 million for the entire San Francisco–Oakland Bay Bridge was just what Fowler had estimated for his design, and the overall project was to include "the greatest bridge yet erected by the human race," according to ex-President Hoover, who spoke at ceremonies marking the beginning of construction in mid-1933. Chief engineer Purcell said he hoped to see traffic using the bridge by the beginning of 1937. In fact, construction would proceed ahead of schedule, and the bridge would be opened in 1936 (see Fig. 9.10), well before the Golden Gate Bridge. In 1955, when it was almost 20 years old, the San Francisco–Oakland Bay Bridge was named by the American Society of Civil Engineers as one of the seven great modern civil engineering wonders of the United States.

BRIDGES AND TRAFFIC

The construction of the Bay Bridge was well documented in photographs, and though the bridge is their clear focus, it is difficult to overlook the large number of ferries that are in the background of so many of the shots of construction workers and the partially completed structure. By the 1930s commuter traffic across the bay had grown to such a volume that

FIGURE 9.10 Aerial view of the entire San Francisco–Oakland Bay Bridge and of Treasure Island, which was created as the site of the 1939 Golden Gate International Exposition commemorating the completion of the Golden Gate and Bay bridges

criss-crossing ferry boats making as many as 500 trips per day presented more of a hazard to navigation than did a limited number of bridge piers in the water. With the completion of the Bay Bridge, the Golden Gate, and other area bridges, ferry service naturally declined, and automobile traffic in and out of San Francisco became a much more convenient experience—until traffic volume grew to such a magnitude that the bridges themselves became choked with rush hour traffic. Building additional bridges today is considered financially prohibitive, not to mention environmentally threatening, and so ferry service has been reintroduced to relieve bridge traffic.

The Golden Gate Bridge, Highway and Transportation District, which developed as an independent entity as a direct result of engineer Strauss's political savvy and promotional skill, has developed alternative and innovative ways to alleviate congestion. For example, in 1968 the Golden Gate became the first bridge to collect tolls in one direction only. Under the assumption that the great bulk of users cross a bridge both ways, doubling the toll and collecting it only one way makes incontrovertible sense, and many toll bridges, including the Bay Bridge, have since adopted this practice. In 1970 the Golden Gate Bridge, Highway and Transportation District introduced a new ferry service between San Francisco and Sausalito, to further reduce traffic and congestion on the bridge.

When the required traffic capacity of the Bay Bridge was considered prior to design, it became clear that a single-deck bridge would have had to be about 100 feet wide, requiring a very strong and stiff floor structure. A double-deck roadway, approximately one third less wide, was found to result in considerable savings, and so it was adopted.

Other factors also affected the final design of the double-deck roadway. For example, as originally conceived, two interurban railway tracks would have been carried over or under the single vehicular roadway, but the double-deck concept made it possible to carry both sets of rapid transit rails more efficiently on the same level as truck traffic. Furthermore, the tracks were laid out side by side on the bridge's lower deck, even though this created the complication of a somewhat asymmetrical load on the structure. This kind of engineering decision was made to take into account more than just structural response to the loads, however. Locating the two rail lines symmetrically on opposite sides of the bridge would have meant that the vehicle traffic between them would have been interrupted and made more dangerous whenever trains had to cross over from one rail line to the other, which would be necessitated by repairs, accidents, and other circumstances. Commuter trains operated on the lower deck of the bridge until the late 1950s, when they were removed because ridership had dropped. The upper and lower roadways were then each dedicated to one-way traffic.

BRIDGES AND EARTHQUAKES

In the San Francisco Bay area a design consideration more critical than the arrangement of traffic lanes and commuter railroad tracks is the possibility of an earthquake moving the ground beneath the foundations of any structure, including a great bridge. If its traffic-carrying capacity is to be maintained under such abnormal circumstances, the bridge must be designed to withstand a certain amount of shaking. But how much extra strength a structure must have to withstand an earthquake is a matter of engineering judgment that must be tempered with economic reality. This means that reasonable assumptions about the size and characteristics of the kind of earthquake that is likely to strike during the lifetime of the bridge must be made. In the case of the Bay Bridge, Chief Engineer Purcell, in an article in *Civil Engineering* in 1934 describing the project when it was under way, revealed how earthquake resistance was built into the design:

> All elements of the bridge are designed for an acceleration of the supporting material of 10 per cent that of gravity. It was readily recognized that the usual criteria for earthquake design would not be satisfactory in dealing with this structure. In view of this, an exhaustive study was made and design methods were evolved which took into consideration the various peculiarities of the problem.
>
> In the case of the channel piers, the horizontal force created by the acceleration of the mass will be augmented by forces due to the movement of the pier through the water and the soft mud immediately below. In fact it is conceivable that the soft mud may have an acceleration of its own in a direction opposite to that of the pier. These forces were incorporated in the analysis. In dealing with the superstructure, and particularly the suspension spans, the elastic and mechanical flexibility of the elements were fully considered. For earthquake design a 40 per cent increase in basic unit stress was permitted.

The same differences in character between the San Francisco–Goat Island crossing and the Goat Island–Oakland crossing that had necessitated two completely different kinds of structural design also turned out to complicate the attempt to design the Bay Bridge to withstand earthquakes. The West Bay section of the bridge would rest on bedrock, while the East Bay section would sit on mud that is vulnerable to seismic movement. It is this mud to which Purcell was referring in his description of the design problem.

In engineering-science courses like strength of materials and dynamics, engineering students learn to calculate the effects of loads and motion on structural elements. Working problems and exercises in these areas prepares engineers for the kinds of calculations alluded to by Purcell. However, unlike in textbook problems, where all the loads and conditions are clearly specified to give the student a well-defined example for analysis, in real design problems like those that faced the engineers of the Bay Bridge, the clearly essential and critical assumptions regarding what the basic accelerations moving the bridge will be and what additional strength it should have to resist them must be made as part of the calculations and design itself.

Because knowledge of the nature of earthquakes is largely empirical and historical, it becomes a matter of judgment as to what conditions will be designed for. As the earthquake of 1989 subsequently demonstrated, the design of the Bay Bridge was in fact not without its limitations, as a section of the upper roadway fell onto the lower, causing death and injury to motorists on the span at the time and closing the bridge to all traffic. The resulting inconvenience to commuters in the 260,000 vehicles that used the bridge each day demonstrated very forcefully not only the key role that the bridge played in bay area traffic patterns but also the vulnerability of all structures to key design decisions made decades earlier. Because the bridge was so important a traffic link across the bay, the repairs were completed with great speed and deliberation. The California Department of Transportation (known as Caltrans), which is responsible for all of California's toll bridges except the Golden Gate Bridge, reopened the Bay Bridge only 30 days after the earthquake. Subsequent analyses of the bridge suggested that the East Bay portion might have to be replaced entirely, and the total cost of all work needed may be as high as $1.3 billion.

By understanding the long history of a project like the San Francisco–Oakland Bay Bridge, not only do we better understand why its behavior in an earthquake is what it is but also we realize that the engineering and other decisions that went into its design and construction are really only fully meaningful in the context of engineering projects that preceded and were contemporary with it. Engineering cannot take place in a social or technical vacuum, and thus the forces that shape any given engineering project are the same forces that shape the other news and world events that are taking place simultaneously. While engineers should never be ignorant of or ignore the technical aspects of engineering that are fundamental to a project and that give it its unique character, these technical aspects make up only part of any engineering problem, which is an invariably complex human endeavor. Every engineering effort is shaped by, and in turn shapes, the culture, politics, and times in which it is embedded.

10

When the architect Le Corbusier called a house "a machine for living in," he drew attention to the fact that modern buildings are more than structures with facades. They are made up of many parts that must fit together and work together in order to provide not only shelter and status but also comfort in a controlled environment. Our homes have heating and cooling systems, and electrical systems to make them function. The larger buildings in which people make travel connections and work have more elaborate systems, which might include those for horizontal and vertical means of transportation comprising many interconnected moving sidewalks, escalators, and elevators. Where government or industrial secrets are involved, and in politically sensitive areas, there may have to be security systems linked to state-of-the-art surveillance equipment.

Though older buildings may not have had the same kinds of systems with which they are now so often retrofitted, that is not to say that older buildings did not also have to be built with systems in mind, for no complex building project can proceed very effectively without them. Not all systems involve mechanical, electrical, and computer hardware, and some of the most effective systems of ancient times were more sociological than technical. The great pyramids of Egypt, for example, required enormous systems of labor organization for their building, which in turn depended upon now incompletely understood systems of moving great quantities of massive blocks of stone. The design of the pyramids also incorporated systems of passages and chambers carefully laid out to thwart would-be tomb robbers.

In later centuries, when giant obelisks were removed from Egypt or relocated in cities like Rome, the operation had to be carefully planned and the muscle power of hundreds, if not thousands, of men and animals had to be coordinated into a smoothly functioning system (Fig. 10.1). While an obelisk itself might be considered a primitive building with no

FIGURE 10.1 Elaborate method devised for moving the Vatican Obelisk

internal systems, if improperly handled and not supported in an appropriate way, it was as liable as any machine to break. Gothic cathedrals were, of course, much more elaborate structural systems, with vaulted ceilings and flying buttresses. However, beyond those dominant mechanical features and the sociological systems that were required for their sometimes centuries-long financing and construction, the great cathedrals did not generally have to meet the tight design, scheduling, or environmental requirements of modern buildings.

THE CRYSTAL PALACE

What might be considered one of the greatest-ever public buildings to require systems thinking for both its construction and its operation was the building designed to house the first World's Fair, officially known as the Great Exhibition of the Works of Industry of All Nations, which was held in London in 1851. The idea to have such an international exhibition developed naturally out of the increasingly inventive spirit and productivity of the Industrial Revolution. By the middle of the nineteenth century, local and national exhibitions of the products of industrial art and manufacture had become commonplace, but it was the vision of the

By the late 1980s, offshore oil platforms standing on the sea floor reached structural heights that made them taller than the highest buildings on land. The Shell Oil Company's Bullwinkle platform, named after the oil field in the Gulf of Mexico that it was designed to exploit, was taller than the Sears Tower, and was named by the American Society of Civil Engineers as the Oustanding Civil Engineering Achievement of 1989. The steel structure weighed more than 75,000 tons and was constructed in a horizontal position and towed in that position to its site, where it was launched from barges to sit in a vertical position in the deep water.

What structural precautions must be taken in moving an off-shore platform from a horizontal to a vertical position? How would those differ from the precautions taken in lifting a stone obelisk to a vertical position?

British to escalate the concept across national boundaries. Prince Albert became a staunch supporter, and with the appointment of a royal commission under his chairmanship the idea was formally launched early in 1850.

The ambitious idea was to construct a temporary building in London's Hyde Park capable of sheltering all the displays and people under one roof that covered an unprecedented sixteen acres. Like many building projects, it began with the invitation of designs for the structure, and 245 entries were received. The committee found none of them to be suitable, however, and so borrowed from them to propose a design of its own. But the committee's massive structure, complete with a dome larger than the great one on London's St. Paul's Cathedral, was judged impractical by critics. This, coupled with growing debate in Parliament over whether the invitation to so many foreigners to bring their goods and visit London would threaten not only the country's balance of trade but also the city's health, did not augur well for the exhibition. In such a climate, the Great Exhibition might never have taken place at all had a viable building plan not surfaced.

Joseph Paxton was the superintendent of the gardens at Chatsworth, the estate of the Duke of Devonshire, and had become well know for his design and construction of an iron and glass greenhouse that covered an acre of ground, thus serving as a model for later palm houses and conservatories. Paxton had also won fame for having designed and constructed at Chatsworth a special building in which a giant tropical water lily, which he named after Queen Victoria, not only bloomed but thrived. Such structures had to be supported by carefully controlled systems that regulated water supply, humidity, heat, and light, and so Paxton had the experience to propose a building for the exhibition that relied upon systems of all kinds and yet would be buildable before the Great Exhibition's scheduled opening. Furthermore, Paxton's building would function well when it was full of people in the heat of the summer. His plan was embraced with less than a year to go before the crowds were to arrive, and the vast wood, iron, and glass building that resulted became known as the Crystal Palace (Fig. 10.2).

The first system that made it all possible was the structure's design and method of construction. The temporary building employed standardized and reusable columns, girders, and roof components, which could be effectively manufactured and assembled. The construction management system was efficient; for example, at the beginning of construction the site

FIGURE 10.2 The Crystal Palace of 1851, erected in London's Hyde Park

was fenced off with the same boards that would in the end be taken down and installed as flooring in the building. A drainage system was laid out, and the building's hollow columns were erected atop it so that they could carry rainwater from the cleverly designed ridge-and-furrow roof directly into the drains. The spacing between all columns was to be either 24 feet or a multiple of that distance, and there were standardized means of connection. Erection was completed in 17 weeks. In essence, the frame building was constructed of prefabricated parts, and the roof could be installed at the same time the walls were being hung. This feature, that the walls were not required to support the building, came to be known as curtain-wall construction, and it is used in all modern skyscrapers.

Constructing the 1850-foot-long Crystal Palace was one thing; having it work as a machine with as many as 90,000 people moving about inside was another. With the structure completed, critics who argued that the building would blow down in the wind or that its galleries would collapse under the weight of crowds were silenced, but it remained to be seen whether the enclosed space would not be as stifling as a hot house. Paxton anticipated such concerns by designing air circulation and heat control systems. Wall panels were louvered for ventilation, and large canvas roof covers were designed to shade the interior. Water evaporating from these even provided a kind of air conditioning. Anticipating cleaning problems, Paxton designed the flooring to have spaces between the boards so that litter could be swept into the crawl space beneath. Since this would have produced a fire hazard, Paxton planned the space to accommodate boys who would collect the litter on a regular basis. With such systems in place, the Crystal Palace and the Great Exhibition that it housed were a tremendous success.

After the exhibition closed, many Londoners wanted to keep the building permanently in Hyde Park, but the agreement had been to restore the park to its prior state. Thus the original Crystal Palace was disassembled, as it had been designed to be, and it was reerected on a still larger scale south of the city, in an area known as Sydenham, where it served for over 80 years as the centerpiece of a recreational and cultural retreat from the city proper. A fire in 1866 destroyed the building's north transept, and in 1936 the building, with its many flammable contents, caught fire and burned down completely. But the influence of the Crystal Palace on engineering, architecture, and building systems generally continues to this day.

TOWERS AND ELEVATORS

Before it was agreed to reconstruct the Crystal Palace at Sydenham, various proposals were put forth for reusing the structural materials. Among them was to construct a tower 1000 feet high, which its proposer pointed out would be an economical use of land and from the top of which the view would no doubt have been spectacular. Though limited space is certainly one of the reasons skyscrapers have come to be so popular in crowed cities, the idea of so tall a tower in the 1850s had the serious drawback of how to transport people from the ground to the top. Anyone who has climbed the stairs to the top of the Statue of Liberty or the Washington Monument, neither of which is nearly so tall, knows that such a proposition would have limited appeal and would not likely have repaid its investment in tourist trade.

The difficulty of getting to the top of tall structures was clearly one factor in keeping them from being built. In the wake of the success of the Great Exhibition, other large cities held their own international exhibitions, often in crystal palaces of modified design. The one in New York in 1853 was in a cruciform iron-and-glass structure topped by a dome 168 feet high. An alternative building proposal for that exhibition had been put forward by the mechanical and structural engineer James Bogardus, who was a proponent of cast-iron buildings. He had wanted to erect a 300-foot-high cast-iron tower that would have transported visitors to the top in a steam elevator. Elevators in use at the time were installed mainly for transporting freight and were operated by hydraulic power, much the way an automobile lift is in a garage today. It was not uncommon for a hydraulic cylinder to fracture or a lifting rope to break on early elevators, and having people free-falling from a few stories, let alone a 300-foot tower, was naturally of some concern.

What changed the perception of elevators was a demonstration by one of Bogardus's contemporaries at the same exhibition that rejected his tower. Elisha Graves Otis was a mechanical engineer who developed a safety device that would check an elevator's fall if the supporting rope or cable broke. At the New York exhibition, Otis had constructed a frame and elevator in which he was hoisted some distance above the ground (Fig. 10.3). When an assistant, in clear view of all onlookers, dramatically cut the supporting rope, the elevator dropped only a small distance before being stopped by the new safety feature. By 1857 a hydraulic passenger elevator was installed in a five-story store on Broadway, and in time the idea of elevators in buildings was commonplace. While Bogardus, Otis, and others had developed steam-powered elevators, in general that power supply was used to drive pumps for hydraulic plunger-type elevators in which a piston was set deep into the ground beneath the elevator shaft. This arrangement clearly imposed practical height limitations. Nevertheless, New York's famous Flatiron Building, with about 20 stories, had hydraulic elevators operating into the last decades of the twentieth century.

With the advent of elevators, the practical limits to the height of a building were no longer defined by how many flights people were able or willing to walk up but rather by the structural limitations of how tall a building could be constructed. The standard methods of construction employed masonry walls as the principal means of support. Since all stone has a limit as to how much weight can be piled upon it before it crushes, taller masonry structures required walls which, at least near ground level, had to be prohibitively thick. The advent of iron-framed structures, such as those pioneered by Paxton and Bogardus, in conjunction with the concept of a curtain wall, removed these limitations and opened up new possibilities. As the end of the nineteenth century approached, the use of cast

FIGURE 10.3 Elisha Otis's famous demonstration of an elevator safety device

FIGURE 10.4 Patent for the Eiffel Tower

iron, wrought iron, and finally steel in construction was becoming more and more common, and so taller and taller structures came to be built.

It was possible in the mid-1880s, for example, to erect the copper shell of the Statue of Liberty to its full 151-foot height atop its pedestal in New York Harbor because it was supported by a wrought iron frame designed by the French bridge engineer Gustav Eiffel. Eiffel is, of course, more commonly associated with the Paris tower that bears his name. Like the Crystal Palace, that tower had its origins in the planning for an international exhibition. The 1889 Exposition Universelle was to be held to commemorate the French Revolution, and a suitably distinctive monument was sought for the occasion. Two engineers working in Eiffel's firm, Maurice Koechlin and Emile Nouguier, came up with the idea of a 300-meter (about 1000-foot) tall tower made of wrought iron. When they first showed their concept to Eiffel, he expressed little enthusiasm. However, after Stephen Sauvestre, an architect with the firm, added some embellishments, Eiffel embraced the idea and eventually a patent on the structural design was secured (Fig. 10.4).

Resisting wind forces was the main structural challenge in designing what came to be known as the Eiffel Tower, but some of the systems that were incorporated into it were also essential for its success. For example, it was clearly imperative that the tower be raised on a true vertical, which taxed the technology of the time. To ensure that things would not get out of alignment, a system of hydraulic jacks was incorporated into the four corner bases, so that corrections could be made as the tower rose. Completing the tower structure would have been a feat in itself, but if people could not be efficiently transported to its top there would be no significant income to repay its cost. Thus, considerable effort went into designing the elaborate elevator system, which had to traverse not only the vertical central spire but also the inclined structural legs that gave the tower its stiffness in the wind. The multiple component double-decker elevator system that moves people to the various levels of the Eiffel Tower remains a masterpiece of planning and was largely responsible for making the tower enough of a profitable venture and popular attraction that it has stood in Paris well beyond its originally intended purpose as a symbol of an exposition.

THE WOOLWORTH BUILDING

The term skyscraper came to be used in the 1880s to describe buildings in Chicago that reached heights of ten to sixteen stories. Though these structures were awe-inspiring in their time, they tend to look squat by to-day's standards, having horizontal dimensions comparable to their heights, and they were nowhere near the height of the Eiffel Tower. In the twentieth century, tall buildings began to rise to much greater heights than their street level width. Among the most notable of the early-twen-tieth-century skyscrapers was the Woolworth Building in New York City. The cost of the 1913 structure was a staggering $13 million, which represented but one of many of the nontechnical obstacles to building skyscrapers. However, in this case the building was paid for in cash by its owner. The strong-willed, self-made Frank W. Woolworth had become wealthy on the basis of his "five and ten cent" retail stores that at the time were doing $100 million in annual sales.

With the financing of the Woolworth Building assured, the detailed problems of design could be faced. Woolworth dictated certain criteria to the architect, Cass Gilbert, for what came to be called the Cathedral of Commerce (Fig. 10.5). The 20-foot-high Gothic ceilings of some floors made the building's 60 stories belie its height of 792 feet above the ground. More traditional ceiling heights would have given the building almost 80 stories. Several of the Woolworth Building's floors were actually below the ground, calling attention to one of the first problems faced in designing such structures: the foundation system. A foundation that will support the heavy weight of the building without uneven settling is essential for a stable structure that will not tilt or crack over time.

The foundations of the Woolworth Building, which was estimated to weigh a total of about 223,000 tons, were constructed through soil, mud, silt, and water to the bedrock that defines Manhattan Island. To reach an average depth of 110 feet below street level, an excavation and construc-tion system employing pneumatic caissons was used. These were large inverted steel cylinders, opened at one end like an empty food can, so that they would trap air much the way an inverted glass does in a tub of water. By adding the right amount of air pressure (the pneumatic part), the water through which the excavation proceeded could be kept at a sufficiently low level so that workmen could move the soil, dirt, and loose rocks to reach the firm bedrock. Such procedures, while potentially dangerous because of the compressed air environment, were familiar in construction, especially in large bridge-building projects.

By contrast, the problems presented by the interior of the Woolworth Building were unique to the twentieth century. These included installing the mechanical and electrical systems needed to make a skyscraper func-

FIGURE 10.5 The Woolworth Building

tion properly. For example, the Woolworth Building, with 87 miles of electrical wiring, was the first to have its own power plant, which consisted of four generators capable of producing enough electricity to satisfy the needs of a small city. As part of the building's opening ceremonies, its 80,000 light bulbs were simultaneously lighted when President Woodrow Wilson threw a switch in the White House. The copper roofs of the building were connected via copper cables to the steel skeleton in order to ground it against lightning strikes.

SKYSCRAPERS AND ELEVATORS

When the Woolworth Building opened in 1913, its elevator system was state of the art. Two of the fastest elevators in the building could cover the 700 feet from the street to the 54th floor in the record time of just one minute. But a total of 26 elevators was needed to serve the 30 acres of office space in the building, and the shafts in which these elevators moved took up valuable floor space. For future skyscrapers, whose financing would depend on how much rental income could be realized from the building, every additional elevator shaft required might threaten the project's future.

The Woolworth Building remained the world's tallest until 1930, when the 1046-foot tall Art Deco-style Chrysler Building was completed. It in turn held the record for only one year. The Empire State Building, like the Crystal Palace, was a model construction project. It took only fourteen months to construct, and yet for over 40 years the 1250-foot tall building (1414 feet when the television tower was added) would hold the record as the tallest building in the world. While the distance to the 102nd-floor observatory was not nearly twice that to the top of the Woolworth Building, the Empire State had almost three times as many elevators. Indeed, the problem of moving people vertically in the tallest buildings has been among the most significant factors in limiting their height. In the mid-1990s, over 60 years after the opening of the Empire State Building, its height remained within 10 percent of that of the world's tallest buildings, the World Trade Center Towers (1368 and 1362 feet tall) in New York and the Sears Tower (1454 feet) in Chicago (Fig. 10.6), completed in 1973 and 1974, respectively.

The fastest elevators in the Empire State Building were about twice as fast as those in the Woolworth Building. By the mid-1990s the fastest elevators in the world, operating at almost 2400 feet/minute, were those installed in the 70-story Landmark Tower in Yokahama, then Japan's tallest

FIGURE 10.6 The Sears Tower

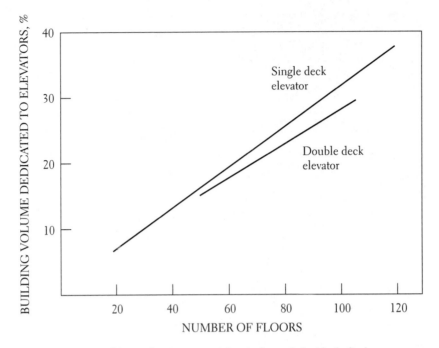

FIGURE 10.7 Building volume occupied by single- and double-deck elevators, as a function of floors in a skyscraper

building. As elevators get faster, however, problems of alignment, noise, and safety grow. Improved roller guides, streamlined and soundproofed cars, and advanced ceramic safety brake shoes are among some of the new features incorporated into the Landmark Tower elevator system.

But speed is only one aspect of the elevator problem. In the typical 100-floor skyscraper, about 30 percent of floor space is devoted to elevators and their appurtenances, such as lobbies and machinery rooms. Some floor space can be saved by employing double-deck elevators (see Fig. 10.7) or operating multiple elevators in a single shaft. No matter how much space it consumes, elevator capacity remains essential not only to move people quickly and conveniently during rush hours but also for evacuating skyscrapers during emergencies, and alternatives to conventional approaches are developing. By 1996 the basic height of the Sears Tower (without its antennas) was surpassed by the 1476-foot tall (about 95-floor) Petronas Twin Towers in Kuala Lumpur, Malaysia (Cesar Pelli, architect; Thornton-Tomasetti, structural engineers). A connecting sky-bridge (Fig. 10.8) enables workers and visitors to move between the towers

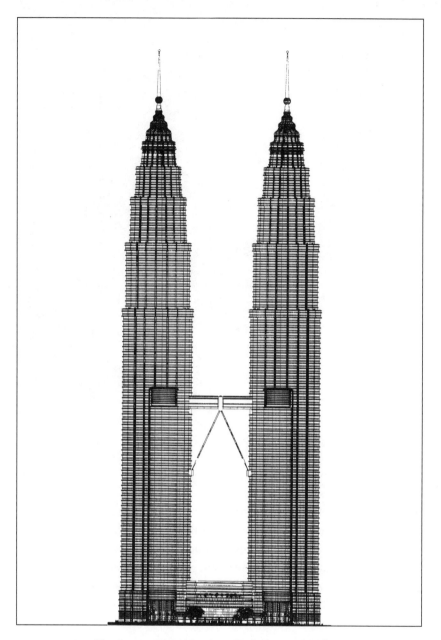

FIGURE 10.8   The Petronas Twin Towers, in Kuala Lumpur, Malaysia

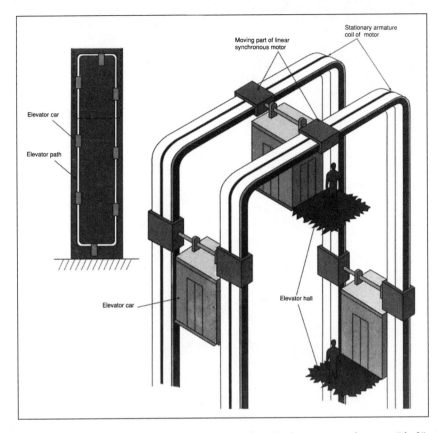

FIGURE 10.9 An advanced elevator system, with multiple cars using the same "shaft" in a Ferris-wheel-like arrangement

without having to take elevators all the way to the ground level, and it also provides an alternate exit in the event of an emergency.

Another concern in elevator design is the length of the ropes and cables supporting the elevator. As they grow longer they begin to add substantial weight to the system. For this and other reasons, elevator engineers are developing elevators in which the cable is eliminated entirely. Among these is a system in which multiple elevator cars are moved in a continuous circuit powered by permanent magnets and linear synchronous motors, as suggested in Fig. 10.9—an idea not unlike that of a Ferris wheel.

MOTION OF TALL BUILDINGS

At the same time that buildings have grown taller, they have grown relatively lighter and hence more flexible. Such evolutionary developments are natural in engineering, and they can be traced in the history of bridges, airplanes, and structures generally. These trends are driven principally by economics, but there are also related aesthetic and functional considerations. A lighter-framed skyscraper, for example, can be sleeker looking architecturally and can have more floor space and window area, which makes it more attractive to potential tenants. However, if the building is too flexible, the perceptible motion of the upper floors can be disconcerting to the occupants, who might see pictures tilt on the walls or coffee slosh around in the cup on their desk.

The design of skyscrapers has evolved with the development of novel structural systems and computer-based analysis. Such concepts did not exist when the Empire State Building was designed, and so it is by today's standards a very massive and stiff structure. Indeed, in 1945 when an airplane crashed full speed into the building during foggy weather, the building suffered relatively minor structural damage. Nevertheless, even the stiffest tall building has a degree of flexibility, and before the spate of tall building construction in the 1960s and 1970s, it was a matter of debate among visitors to the observation deck as to whether the top of the Empire State Building really could sway a foot or two in the wind.

The John Hancock Tower on Copley Square in Boston is not nearly as tall as the Empire State Building, but it has an unusual floor plan that made it susceptible to being twisted by the wind, something that was not anticipated by the architect, Henry Cobb of I. M. Pei & Partners, or the structural designers. At first, the repeated twisting and other building motion was thought to have caused the building's large window panes to fall out onto the square below, posing a considerable safety hazard. But the heating and cooling of the glass itself was found to be the culprit. After the building's problems were properly diagnosed, the windows were redesigned, and the twisting was dealt with by installing on the 58th floor a system of computer-controlled hydraulic actuators attached to two 300-ton blocks of dead weight that could counter any undesired motion in much the same way we can check the movement of a playground swing by throwing our body weight around in a contrary fashion. The system installed in the John Hancock Tower is known as a tuned mass damper (Fig. 10.10), a system that has been installed in other tall buildings as part of their design. When a certain degree of motion is detected by sensors, a pump is activated to float a massive block on a film of oil, thus allowing it to slide relative to the building in accordance with the motion of the programmed actuators to which it is connected.

The first tall building to have a tuned mass damper as part of its design

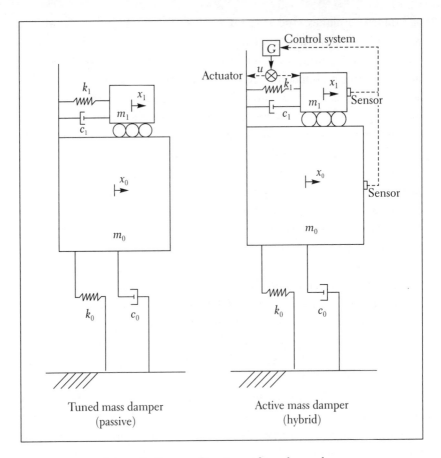

FIGURE 10.10 Schematic diagram of two types of tuned mass damper

proper was the Citicorp Center, completed in 1977 in New York City. After it was decided to build this corporate tower between Fifty-third and Fifty-fourth Streets, directly across Lexington Avenue from Citicorp's headquarters, it was found that a turn-of-the-century church located on one corner had to be accommodated. In exchange for the air rights to build above it, the old church was to be torn down and a new one erected in its place as part of the Citicorp Center complex. The 914-foot tall skyscraper itself was designed by the architect Hugh Stubbins to rest on four columns that rose up nine stories from the street. The tower was cantilevered out over the four open corners, into one of which was tucked the church, to rise another 50 stories with a square floor plan. The top of the building was given a distinctive slanted profile. Because of the unusual shape of its structural skeleton (Fig. 10.11), the Citicorp Tower was de-

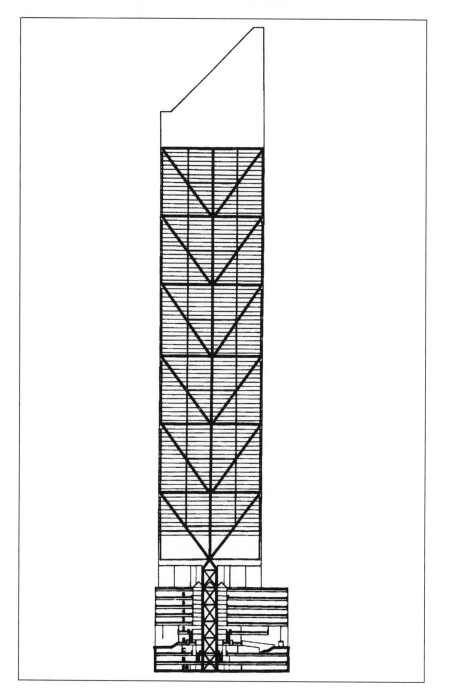

FIGURE 10.11   The structural system of Citicorp Tower

signed from the beginning to contain a 400-ton tuned mass damper to reduce motion induced by the wind, and this inspired the system installed in the Hancock Tower in Boston.

The main resistance against the wind comes from a skyscraper's structural system itself, of course, and designing Citicorp's was especially tricky. The consulting structural engineer for the project was William LeMessurier, whose firm also designed the tuned mass damper in Boston's Hancock building. LeMessurier devised a unique system of cross bracing to transfer the gravity and wind loads from the tower to the four columns on which it rested (Fig. 10.11). The main steel components of the building were to be welded together to form a rigid frame, but as LeMessurier discovered after the building was completed and occupied, bolts had been used instead of the more expensive welding. Such design changes are common in the course of taking a project from conceptual design to construction, and the substitution of bolts for welding was approved by LeMessurier's office. However, because of the unusual way in which the Citicorp structure had to resist the wind, LeMessieur later found that the bolts had not been designed for the worst-case condition, and elaborate emergency welding had to be done before the building was subjected to dangerous winds. The story of the retrofitting was largely unknown to the public until 1995, when it was told in a *New Yorker* magazine story.

UNEXPECTED PROBLEMS

A tube derives its resistance to bending from the fact that the main supporting material is concentrated around the periphery rather than at the center of the structure. This principle lies behind the design of the diagonally braced John Hancock Center in Chicago (Fig. 10.12), completed in 1969. This landmark building was conceived by the structural engineer Fazlur Khan in conjunction with the architect Bruce Graham and engineers and architects in the firm of Skidmore, Owings & Merrill. The Sears Tower (Fig. 10.6), also designed by Graham and Khan and completed five years later, carried the concept a step further by employing nine bundled tubes, each reinforcing the resistance of the others to bending in the wind. Many very tall building designs that have been subsequently proposed have made use of variations of the tube or bundled tube concept.

The twin towers of the World Trade Center (Fig. 10.13) were designed for the Port of New York Authority by the architect Minoru Yamasaki and

FIGURE 10.12 The John Hancock Center, Chicago

FIGURE 10.13 The World Trade Center Towers

the engineering firm of Skilling Helle Christiansen & Robertson. Completed in 1973, the towers house remarkably versatile office space because of the tube concept. In each of the 209-foot square towers the 60-foot-long floor beams span the distance from the columns of the external wall to the stiff elevator core, thus providing an open space that is very flexible. To move the enormous numbers of people in and out and up and down the towers, 230 passenger elevators were designed to operate in a system of lower-, middle-, and upper-floor zones interconnected by skylobbies where people would transfer from one set of elevators to another.

Wind loads as high as 45 pounds per square foot were considered in the structural design of the World Trade Center towers. As has become standard with tall buildings, model tests were conducted in a wind tunnel in order to confirm theoretical and computer calculations. There were also considerations of how the wind channeled to blow between the towers might affect their behavior. The design proved to be remarkably successful, but when an excessively strong hurricane threatened New York City about 20 years after the World Trade Center had been opened, promising to subject the towers to more wind force than they had yet experienced, they were evacuated as a precaution. The hurricane did not strike with its full force, but as it turned out the strength of the towers would be tested in an even less predictable way.

As with skyscrapers generally, there were numerous sub-basement levels beneath the World Trade Center to accommodate mechanical equipment and parking for the cars of the many workers in and visitors to the towers. Shortly after noon on Friday, February 26, 1993, a tremendous explosion in the underground garage caused floors to collapse and smoke to rise in the soon darkened and powerless north tower. A truck full of explosives had been parked near some of the outside columns of the structure, apparently in the hope that the blast would bring down the second tallest building in the world, and bring attention and satisfaction to the terrorist group responsible. While the tower structure itself proved able to resist the attack, the collapse of the sub-basement floors left some columns dangerously unbraced, and shoring them up became the first order of business as soon as the damage was assessed (Fig. 10.14).

Before the structural damage could be fully determined, however, there were tens of thousands of occupants to evacuate. With the power out and the elevators immobilized, some people had to make their way down scores of flights of darkened, smoke-filled stairways. Others were trapped in elevators for hours, and some of them were filling up with smoke rising

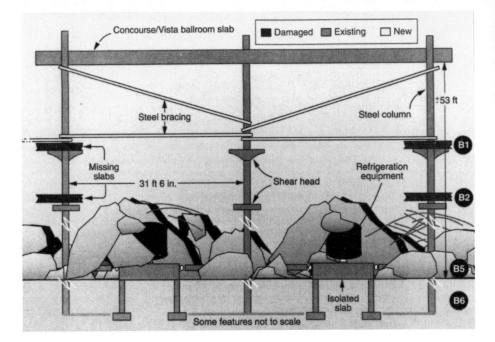

FIGURE 10.14 Terrorist-bomb damage to the World Trade Center, showing the emergency bracing of columns

in the shafts. Among those trapped in elevators were some engineers associated with the Port Authority of New York and New Jersey, the present name of the owner of the World Trade Center as well as of many of the bridges that could be seen from its offices high in the north tower. One elevator that was stuck near the 58th Floor held the Port Authority's chief engineer, Eugene Fasullo, who knew the building inside and out because he had worked on its design. After prying open the elevator doors, the trapped occupants found themselves facing a blank wall. Fasullo knew that cutting through the heavy-duty wall board would take them into some public space from which they could reach stairs, and so they scratched, scraped, and dug with the few implements they had available — keys, paper clips, nail cutters, and the panel from the elevator's console. After three hours, the group cut a hole large enough to crawl through to a restroom and thus escaped to safety.

Upon reaching the ground, Chief Engineer Fasullo immediately began to help assess the damage, and he soon became somewhat of a celebrity not only because of his determined escape but also because of his knowledgeable position with regard to the condition of the building and its systems. His forthrightness and articulateness during numerous news conferences in subsequent days and weeks kept the public informed of the nature of the damage and the progress of its repair. Due in large part to Fasullo and his colleagues, the World Trade Center was repaired and reopened in a remarkably short time, but with much enhanced security and much restricted access to underground spaces.

Terrorism is a very difficult thing to design against, in part because a particular size bomb blast must be chosen as a "design blast," and no matter how large the most credible blast might be, an even larger one might become feasible after a structure is built. Natural events like hurricanes and earthquakes can also only be designed against to a degree. The historical record over a period of, say, fifty or a hundred years may serve as the basis for establishing what constitutes the worst storm to expect during the anticipated lifetime of a building. This may serve as a design basis, but there will always be the outside chance that a greater storm will occur. Earthquakes present even more critical design decisions, and experience has shown that although past earthquakes in a region had certain characteristics that could serve as a basis for rational design of structures like skyscrapers and bridges, the earth still moves in mysterious ways and holds many surprises as to the magnitude and direction of those movements. The Northridge Earthquake that struck the Los Angeles area in 1994, for example, occurred along a previously unknown geological fault and produced uncharacteristically large vertical ground motions. This, in part, accounted for the unexpectedly large amount of damage to bridges and steel-framed buildings, which had therefore generally been thought to resist earthquakes better than masonry and concrete ones.

ENVIRONMENTAL FACTORS

Though terrorist blasts and natural disasters might test a large building to its limits, more everyday considerations can often determine its long-term success or failure as a complex system comprising many subsystems. Climate control inside a modern building is very important, and the development of air conditioning has often been given much credit for the commercial growth of such famously hot and humid cities as Atlanta and Houston. As early as the 1920s, theaters were air-conditioned, but highrise buildings presented a problem because of the large volume that had to be cooled and the numerous bulky air ducts that were required to achieve the volume of air flow needed at the upper floors. The first highrise building to be air-conditioned effectively was the 21-story Milam Building in San Antonio, Texas, in which the Carrier Engineering Corporation installed a 300-ton refrigeration unit in 1928. Cooled water was pumped throughout the building to where a series of fans blew warm air over the cool pipes. Water from the San Antonio River was used to cool the system's condenser until the practice was stopped in the 1950s, and a cooling tower had to be installed.

Large cooling towers are now commonplace on large buildings, but when improperly operated they have been the source of confounding disease. One such mysterious outbreak occurred during an American Legion convention, hence the name Legionnaire's Disease. Other unexpected environmental complications of modern buildings have developed in offices whose windows were permanently sealed so that the heating and cooling systems could better control the climate inside. This led to improper ventilation, in some cases combined with toxic fumes given off by synthetic materials and solvents, that have given the occupants headaches, nausea, and skin rashes. The final outcome in a few cases has been an unoccupiable "sick building."

With the completion of more and more very tall buildings holding as many occupants as there are in a good-sized city, it has also become increasingly clear that huge buildings have a profound environmental impact on the community in which they are located. The sheer volume of people and the services they demand can tax all sorts of systems, including transportation, mail, telephone, and public health. The World Trade Center, for example, must accommodate about 50,000 employees and 80,000 visitors daily, which at any given time give it the population of a major city and all the attendant complexities. As with all engineering designs, the successful building is one in which all sorts of complications and interactions of the main system and its various subsystems, and their impact on and interaction with the environs and its systems, are properly anticipated by their engineers.

Books listed were generally first published in the year given directly after the author's name. Where subsequent editions, translations, or other referenced versions are also available, their date of publication is given at the end of an entry.

1. Introduction

Adams, James L. 1991. *Flying Buttresses, Entropy, and O-Rings: The World of an Engineer*. Cambridge, Mass.: Harvard University Press.

Armytage, W. H. G. 1976. *A Social History of Engineering*. Boulder, Colo.: Westview Press.

de Camp, L. Sprague. 1963. *The Ancient Engineers*. New York: Doubleday.

Chrimes, Mike. 1991. *Civil Engineering, 1839–1889: A Photographic History*. London: Thomas Telford.

Ferguson, Eugene S. 1992. *Engineering and the Mind's Eye*. Cambridge, Mass.: MIT Press.

Finch, James Kip. 1960. *The Story of Engineering*. New York: Anchor Books.

Florman, Samuel. 1976. *The Existential Pleasures of Engineering*. New York: St. Martin's Press.

Fraser, Chelsea. 1928. *The Story of Engineering in America*. New York: Thomas Y. Crowell.

Fredrich, Augustine J., ed. 1989. *Sons of Martha: Civil Engineering Readings in Modern Literature*. New York: American Society of Civil Engineers.

Gille, Bertrand. 1966. *Engineers of the Renaissance*. London: Cambridge, Mass.: MIT Press.

Grayson, Lawrence P. 1993. *The Making of an Engineer: An Illustrated History of Engineering Education in the United States and Canada*. New York: John Wiley & Sons.

Hall, Donald. 1984. *A History of Engineering in Classical and Medieval Times*. La Salle, Ill.: Open Court.

Hapgood, Fred. 1993. *The Infinite Corridor: MIT and the Technical Imagination*. Reading, Mass.: Addison-Wesley.

Hodges, Henry. 1992. *Technology in the Ancient World*. New York: Barnes & Noble Books.

Kidder, Tracy. 1981. *The Soul of a New Machine*. Boston: Little, Brown.

Kirby, Richard Shelton, et al. 1956. *Engineering in History*. Reprint edition. New York: Dover Publications, 1990.

Landels, J. G. 1978. *Engineering in the Ancient World*. Berkeley: University of California Press.

Ley, Willy. 1960. *Engineers' Dreams*. New York: Viking.

Meehan, Richard L. 1981. *Getting Sued and Other Tales of the Engineering Life*. Cambridge, Mass.: MIT Press.

Pannell, J. P. M. 1977. *Man the Builder: An Illustrated History of Engineering*. New York: Crescent Books.

Petroski, Henry. *Design Paradigms: Case Histories of Error and Judgment in Engineering*. New York: Cambridge University Press.

Rae, John, and Rudi Volti. 1993. *The Engineer in History*. New York: Peter Lang.

Reynolds, Terry, ed. 1991. *The Engineer in America: A Historical Anthology from Technology and Culture*. Chicago: University of Chicago Press.

Rolt, L. T. C. 1974. *Victorian Engineering*. Hammondsworth, Middlesex: Penguin Books.

Schodek, Daniel L. 1987. *Landmarks in American Civil Engineering*. Cambridge, Mass.: MIT Press.

Smiles, Samuel. 1874. *Lives of the Engineers*. Popular edition in five volumes. London: John Murray, 1904.

Straub, Hans. 1952. *A History of Civil Engineering*. Translated by Erwin Rockwell. Cambridge, Mass.: MIT Press, 1964.

White, Pepper. 1991. *The Idea Factory: Learning to Think at MIT*. New York: Dutton.

2. Paper Clips and Design

Aristotle. 4th c. B.C. *Minor Works*. Translated by W. S. Hett. Cambridge, Mass.: Harvard University Press, 1980.

Brown, Kenneth A. 1988. *Inventors at Work: Interviews with 16 Notable American Inventors*. Redmond, Wash.: Microsoft Press.

Brunel, Isambard. 1870. *The Life of Isambard Kingdom Brunel: Civil Engineer*. London: Longmans, Green.

Carr, Fred K. 1995. *Patents Handbook: A Guide for Inventors and Researchers to Searching Patent Documents and Preparing and Making an Application*. Jefferson, N.C.: McFarland and Co.

Cooper-Hewitt Museum. 1984. *American Enterprise: Nineteenth-Century Patent Models*. New York: Cooper-Hewitt Museum.

Edwards, Owen. 1989. *Elegant Solutions: Quintessential Technology for a User-Friendly World*. New York: Crown.

French, M. J. 1985. *Conceptual Design for Engineers*. Second edition. London/Berlin: The Design Council/Springer-Verlag.

Goldberger, Paul. 1985. *On the Rise: Architecture and Design in a Postmodern Age*. New York: Penguin Books.

Hollins, Bill, and Stuart Pugh. 1990. *Successful Product Design: What to Do and When*. London: Butterworths.

Institute for Electrical and Electronics Engineers. 1979. Special issue on patents. *IEEE Transactions on Professional Communication*, Vol. PC-22, No. 2.

Leonhardt, Fritz. 1984. *Bridges: Aesthetics and Design*. Cambridge, Mass.: MIT Press.

MacLeod, Christine. 1988. *Inventing the Industrial Revolution: The English Patent System, 1660–1800*. Cambridge: Cambridge University Press.

Merges, Robert Patrick. 1992. *Patent Law and Policy: Cases and Materials*. Charlottesville, Va.: Michie.

Naj, Amal Kumar. 1995. "Hey, Get a Grip! Your Basic Paper Clip Is Like a Mousetrap." *The Wall Street Journal*, July 24, pp. A1, A5.

Pressman, David. 1991. *Patent It Yourself*. Third edition. Berkeley, Calif.: Nolo Press.

Petroski, Henry. 1992. *The Evolution of Useful Things*. New York: Alfred A. Knopf.

——— 1992. "The Evolution of Artifacts." *American Scientist*, September-October, pp. 416–420.

——— 1995. "Learning from Paper Clips." *American Scientist*, July-August, pp. 313–316.

Pugh, Stuart. 1991. *Total Design: Integrated Methods for Successful Product Engineering*. Reading, Mass.: Addison-Wesley.

Riordan, Teresa. 1994. "Patents." *New York Times*, September 19, p. C2.

Rolt, L. T. C. 1957. *Isambard Kingdom Brunel*. Harmondsworth, Middlesex: Penguin Books, 1970.

Timoshenko, Stephen. 1953. *History of Strength of Materials: With a Brief Account of the History of Theory of Elasticity and Theory of Structures*. New York: Dover Publications, 1983.

U.S. Patent Office. 1888. *Women Inventors to Whom Patents Have Been Granted by the United States Government, 1790 to July 1, 1888*. Washington, D.C.: Government Printing Office. (See also supplements issued in 1892 and 1895.)

3. Pencil Points and Analysis

Agricola, Georgius. 1556. *De Re Metallica*. Translated by Henry Clark Hoover and Lou Henry Hoover. New York: Dover Publications, 1950.

Cowin, S. C. 1983. "A Note on Broken Pencil Points." *Journal of Applied Mechanics* 50: 453–454.

Cronquist, D. 1979. "Broken-off Pencil Points." *American Journal of Physics* 47: 653–655.

Galileo. 1638. *Dialogues Concerning Two New Sciences*. Translated by Henry Crew and Alfonso de Salvio. New York: Dover Publications, 1954.

Gordon, J. E. 1978. *Structures: Or, Why Things Don't Fall Down*. New York: Da Capo Press.

——— 1976. *The New Science of Strong Materials: Or Why You Don't Fall*

Through the Floor. Second Edition. Princeton, N.J.: Princeton University Press, 1984.

Petroski, Henry. 1987. "On the Fracture of Pencil Points." *Journal of Applied Mechanics* 54: 730–733.

——— 1990. *The Pencil: A History of Design and Circumstance.* New York: Alfred A. Knopf.

Walker, Jearl. 1979. "The Amateur Scientist." *Scientific American,* February, pp. 158–166. (See also November, pp. 202–204.)

4. Zippers and Development

Ausnit, Steven, et al. 1964–1984. U.S. Patents Nos. 3,160,934; 3,172,443; 3,173,184; 3,203,062; 3,220,076; 3,338,284; 3,347,298; 4,479,244.

Federico, P. J. 1946. "The Invention and Introduction of the Zipper." *Journal of the Patent Office Society* 28: 855–876.

French, Michael. 1994. *Invention and Evolution: Design in Nature and Engineering.* Second edition. Cambridge: Cambridge University Press.

Friedel, Robert. 1994. *Zipper: An Exploration in Novelty.* New York: W. W. Norton.

Gray, James. 1963. *Talon, Inc.: A Romance of Achievement.* Meadville, Pa.: Talon, Inc.

Macdonald, Anne L. 1992. *Feminine Ingenuity: Women and Invention in America.* New York: Ballantine Books.

Naito, Kakuji. 1965–1967. U.S. Patents No. 3,198,228; 3,291,177; 3,340,116.

Panati, Charles. 1987. *Extraordinary Origins of Everyday Things.* New York: Harper & Row.

Petroski, Henry. 1992. *The Evolution of Useful Things.* New York: Alfred A. Knopf.

——— 1993. "On Dating Inventions." *American Scientist,* July-August, pp. 314–318.

Rabinow, Jacob. 1990. *Inventing for Fun and Profit.* San Francisco: San Francisco Press.

Stanley, Autumn. 1993. *Mothers and Daughters of Invention: Notes for a Revised History of Technology.* New Brunswick, N.J.: Rutgers University Press, 1995.

U.S. Patent and Trademark Office. 1990. *Buttons to Biotech: U.S. Patenting by Women, 1977 to 1988.* Washington, D.C.: U.S. Department of Commerce. (See also Update Supplement to 1989.)

Walsh, George M. 1986. "Gripping Saga." *Gannett Westchester* (New York) *Newspapers.* December 21, pp. H1–H2.

Weiner, Lewis. 1983. "The Slide Fastener." *Scientific American,* June, pp. 132–136, 138, 143–144.

5. Aluminum Cans and Failure

Alexander, Christopher. 1964. *Notes on the Synthesis of Form.* Cambridge, Mass.: Harvard University Press.

Bell, R. A. 1965–1966 "Origins of the Canning Industry." *Transactions of the Newcomen Society* 38: 145–151.

Church, Fred L. 1992. "Lightweighting the Aluminum Can: Ain't Over 'til It's Over." *Modern Metals*, December, pp. 34V–34X.

———— 1993. "Fluted Design Helps Cut Beverage Can Weight 10%." *Modern Metals*, February, pp. 34F–34K.

———— 1993. "Steel Can Readied for Beverage Challenge." *Modern Metals*, June, pp. 34FF–34HH.

Collins, Glenn. 1995. "Ten Years Later, Coca-Cola Laughs at 'New Coke'." *New York Times*, April 11, Business Section.

Cowan, Ruth Schwartz. 1983. *More Work for Mother: The Ironies of Household Technology from the Open Hearth to the Microwave.* New York: Basic Books.

Davis, Julie. 1993. "Can Recycling Hits Record Highs in '92." *Modern Metals*, June, pp. 34BB-34EE.

Eagar, Thomas W. 1995. "Bringing New Materials to Market." *Technology Review*, February/March, pp. 43–49.

Hosford, William F., and John L. Duncan. 1994. "The Aluminum Beverage Can." *Scientific American*, September, pp. 48–53.

Lave, Lester, et al. 1995. "My Shopping Trip with André." *Technology Review*, February/March, pp. 59–63.

Myers, John. 1994. "PET Bottle Growth Is Outpacing Resins Supply." *Modern Plastics*, December, pp. 40–45.

Petroski, Henry. 1992. "Form Follows Failure." *American Heritage of Invention and Technology*, Fall, pp. 54–61.

———— 1985. *To Engineer Is Human: The Role of Failure in Successful Design.* New York: Vintage Books, 1992.

Reynolds Metals Company v. The Continental Group, Inc. No. 76 C 4198. United States District Court, N. D. Illinois, E. D. July 6, 1981.

6. Facsimile and Networks

Cole, Bernard C. 1990. "Chip Makers Eye a New Market: PC Fax." *Electronics*, April, pp. 72–74.

Coopersmith, Jonathan. 1993. "Facsimile's False Starts." *IEEE Spectrum*, February, pp. 46–49.

Costigan, Daniel M. 1978. *Electronic Delivery of Documents and Graphics.* New York: Van Nostrand Reinhold.

Frankel, David. 1995. "ISDN Reaches the Market." *IEEE Spectrum*, June, pp. 20–25.

Gawron, L. J. 1991. "Scanned-Image Technologies Bring New Ways to Conduct Business." *AT&T Technologies*, vol 6., no. 4, pp. 2–9.

Honda, Toyota, et al. 1992. "Compacting Technologies for Small Size Personal Facsimile." *IEEE Transactions on Computer Electronics* 38: 417–423.

Hughes, Thomas P. 1983. *Networks of Power: Electrification in Western Society, 1880–1930.* Baltimore: Johns Hopkins University Press.

———— 1989. *American Genesis: A Century of Invention and Technological Enthusiasm, 1870–1970.* New York: Viking.

Hunter, R., I. Dixon, and W. Ablard. 1994. "Fax-on-Demand Systems for Business and Home Applications." *BT Technology Journal* 12: 34–43.

Jordahl, Gregory. 1989. "Instant Information for the 'Need It Yesterday' World." *Manufacturing Systems*, February, pp. 46–50.

Leonard, Milt. 1991. "Fax Modem Chips Bring Datacom to Laptops." *Electronic Design*, March 28, pp. 47–54.

Licwinko, J. S., and F. X. Lukas. 1989. "AT&T Fax Products and Services Speed the Written Message." *AT&T Technology*, vol. 4, no. 2, pp. 12–17.

Lyons, Richard D. 1993. "Austin Cooley, 93, an Executive and a Developer of Fax Machine." Obituary. *New York Times*, September 9.

McConnell, Kenneth R., Dennis Bodson, and Richard Schaphorst. 1989. *Fax: Digital Facsimile Technology and Applications.* Norwood, Mass.: Artech House.

Miastkowski, Stan. 1994. "WinFax Pro Hits the Network." *Byte*, February, pp. 141–144.

Milewski, Allen E., and Henry S. Baird. 1990. "A FAX Reader for the Blind." *Conference Record, Twenty-Fourth Asilomar Conference on Signals, Systems & Computers*, vol. 2, pp. 881–886.

Norford, Leslie K., and Cyane B. Dandridge. 1993. "Near-term Technology Review of Electronic Office Equipment." *Conference Record, IEEE Industry Applications Conference*, Part II, pp. 1355–1362.

Pettersson, Gunnar. 1995. "ISDN: From Custom to Commodity Service." *IEEE Spectrum*, June, pp. 26–31.

Strauss, Neil. 1995. "Pennies that Add Up to $16.98: Why CD's Cost So Much." *New York Times*, July 5, pp. B1, B6.

7. Airplanes and Computers

Bowers, Peter. M. 1989. *Boeing Aircraft: Since 1916.* Annapolis, Md.: Naval Institute Press.

Casey, Steven. 1993. *Set Phasers on Stun: And Other True Tales of Design, Technology, and Human Error.* Santa Barbara, Calif.: Aegean.

Del Valle, Christina, and Michael Schroeder. 1996. "Did the FAA Go Easy on Boeing?" *Business Week*, January 29, pp. 56–58, 60.

Fielder, John H., and Douglas Birsch. 1992. *The DC-10 Case: A Study in Applied Ethics, Technology, and Society.* Albany: State University of New York Press.

Gordon, J. E. 1978. *Structures: Or Why Things Don't Fall Down.* New York: Da Capo Press.

Gottschalk, Mark A. 1994. "How Boeing Got to 777th Heaven." *Design News*, September 12, pp. 50–56.

Lammers, Susan. 1986. *Programmers at Work: Interviews.* First series. Redmond, Wash.: Microsoft Press.

McIntyre, Ian. 1992. *Dogfight: The Transatlantic Battle Over Airbus.* Westport, Conn.: Praeger.

Neumann, Peter G. 1995. *Computer-Related Risks*. New York & Reading, Mass.: ACM Press & Addison-Wesley.

Norman, Donald A. 1989. *The Design of Everyday Things*. New York: Doubleday.

———— 1992. *Turn Signals Are the Facial Expressions of Automobiles*. Reading, Mass.: Addison-Wesley.

O'Connor, Leo. 1995. "Keeping Things Moving at Denver International Airport." *Mechanical Engineering*, pp. 90–93.

Orlebar, Christoper. 1986. *The Concorde Story: Ten Years of Service*. Twickenham, Middlesex: Temple Press.

Petroski, Henry. 1994. "The Draper Prize." *American Scientist*, March–April, pp. 114–117.

Rich, Ben, and Leo Janos. 1994. *Skunk Works: A Personal Memoir of My Years at Lockheed*. Boston: Little, Brown.

Sabbagh, Karl. 1996. *21st-Century Jet: The Making and Marketing of the Boeing 777*. New York: Schribner.

Schlager, Neil, ed. 1994. *When Technology Fails: Significant Technological Disasters, Accidents, and Failures of the Twentieth Century*. Detroit: Gale Research.

Serling, Robert J. 1992. *Legend and Legacy: The Story of Boeing and Its People*. New York: St. Martin's Press.

Vincenti, Walter G. 1990. *What Engineers Know and How They Know It: Analytical Studies from Aeronautical History*. Baltimore, Md.: Johns Hopkins University Press.

Wiener, Lauren Ruth. 1993. *Digital Woes: Why We Should Not Depend on Software*. Reading, Mass.: Addison-Wesley.

Zachary, G. Pascal. 1994. *Showstopper! The Breakneck Race to Create Windows NT and the Next Generation at Microsoft*. New York: Free Press.

8. Water and Society

American Water Works Association. 1990. *Water Quality and Treatment: A Handbook of Community Water Supplies*. Fourth edition. New York: McGraw-Hill.

Carson, Rachel. 1962. *Silent Spring*. Boston: Houghton Mifflin.

Davis, Margaret Leslie. 1993. *William Mulholland and the Inventing of Los Angeles*. New York: HarperCollins.

Fowler, David. 1991. "Drain Brain" [on Sir Joseph Bazalgette]. *New Civil Engineer*, March 14, pp. 24–27.

Frontinus. 97 A.D. *The Two Books on the Water Supply of the City of Rome*. Translated by Clemens Herschel. Boston: New England Water Works Association, 1973.

Kirby, Richard Shelton, et al. 1956. *Engineering in History*. New York: Dover Publications, 1990.

Levi, Enzo. 1995. *The Science of Water: The Foundation of Modern Hydraulics*.

Translated by Daniel E. Medina. New York: American Society of Civil Engineers.

Marco, Gino J., Robert M. Hollingworth, and William Durham, eds. 1987. *Silent Spring Revisited.* Washington, D.C.: American Chemical Society.

Martin, Edward J., and Edward T. Martin. 1991. *Technologies for Small Water and Wastewater Systems.* New York: Van Nostrand Reinhold.

McCullough, David. 1977. *The Path Between the Seas: The Creation of the Panama Canal, 1870–1914.* New York: Simon and Schuster.

Metcalf, Leonard, and Harrison P. Eddy. 1928. *American Sewerage Practice: Volume I. Design of Sewers.* New York: McGraw-Hill.

Moeller, Beverley Bowen. 1971. *Phil Swing and Boulder Dam.* Berkeley: University of California Press.

Morgan, Arthur E. 1971. *Dams and Other Disasters: A Century of the Army Corps of Engineers in Civil Works.* Boston: Porter Sargent.

Nielsen, David M., ed. 1991. *Practical Handbook of Ground-Water Monitoring.* Chalsea, Mich.: Lewis Publishers.

Reisner, Marc. 1986. *Cadillac Desert: The American West and Its Disappearing Water.* New York: Viking.

Robins, F. W. 1946. *The Story of Water.* London: Oxford University Press.

Shallat, Todd. 1994. *Structures in the Stream: Water, Science, and the Rise of the U.S. Army Corps of Engineers.* Austin: University of Texas Press.

Shuriman, Gerard, and James E. Slossom. 1992. *Forensic Engineering: Environmental Case Histories for Civil Engineers and Geologists.* New York: Academic Press.

Smith, Norman. 1971. *A History of Dams.* London: Peter Davies.

Steel, Ernest W. 1960. *Water Supply and Sewerage.* Fourth edition. New York: McGraw-Hill.

Stevens, Joseph E. 1988. *Hoover Dam: An American Adventure.* Norman: University of Oklahoma Press.

Tarr, Joel A., and Gabriel Dupuy, eds. 1988. *Technology and the Rise of the Networked City in Europe and America.* Philadelphia: Temple University Press.

Viessman, Warren, Jr., and Mark J. Hammer. 1985. *Water Supply and Pollution Control.* Fourth Edition. New York: Harper & Row.

Vitruvius. 1st cent. B.C. *The Ten Books on Architecture.* Translated by M. H. Morgan, 1914. New York: Dover Publications, 1960.

9. Bridges and Politics

Ammann, O. H. 1918. "The Hell Gate Arch Bridge and the Approaches of the New York Connecting Railroad over the East River in New York City." *Transactions of the American Society of Civil Engineers* 82: 852–1004.

———— 1933. "George Washington Bridge: General Conception and Development of Design." *Transactions of the American Society of Civil Engineers* 97: 1–65.

Anderson, Graham, and Ben Roskrow. 1994. *The Channel Tunnel Story.* London: E. & F. N. Spon.

Billington, David P. 1979. *Robert Maillart's Bridges: The Art of Engineering.* Princeton, N. J.: Princeton University Press.

———— 1983. *The Tower and the Bridge: The New Art of Structural Engineering.* Princeton, N. J.: Princeton University Press, 1985.

California Toll Bridge Authority. 1937. *Fourth Annual Progress Report.*

Davidson, Frank P., with John Stuart Cox. 1993. *Macro: A Clear Vision of How Science and Technology Will Shape Our Future.* New York: Morrow.

Davidson, Frank P., and C. Lawrence Meador, eds. 1992. *Macro-Engineering: Global Infrastructure Solutions.* New York: Ellis Horwood.

DeLony, Eric. 1993. *Landmark American Bridges.* New York: American Society of Civil Engineers.

Doig, Jameson W., and David P. Billington. 1994. "Ammann's First Bridge: A Study in Engineering, Politics, and Entrepreneurial Behavior." *Technology and Culture* 35: 537–570.

Fowler, Charles Evan. 1915. *The San Francisco—Oakland Cantilever Bridge.* New York: privately printed.

Hopkins, H. J. 1970. *A Span of Bridges: An Illustrated History.* New York: Praeger.

Hunt, Donald. 1994. *The Tunnel: The Story of the Channel Tunnel 1802–1894.* Upton upon Severn, Worcestershire: Images.

Jackson, Donald C. 1988. *Great American Bridges and Dams.* Washington, D.C.: Preservation Press.

Kline, Ronald L. 1992. *Steinmetz: Engineer and Socialist.* Baltimore: Johns Hopkins University Press.

Koerte, Arnold. 1992. *Two Railway Bridges of an Era: Firth of Forth and Firth of Tay.* Basel: Birkhauser.

Mackay, Shelia. 1990. *The Forth Bridge: A Picture History.* Edinburgh: Moubray House.

Mark, Hans. 1987. *The Space Station: A Personal Journey.* Durham, N. C.: Duke University Press.

McCullough, David. 1982. *The Great Bridge.* New York: Simon and Schuster.

———— 1977. *The Path Between the Seas: The Creation of the Panama Canal, 1870–1914.* New York: Simon and Schuster.

Paxton, Roland, ed. 1990. *100 Years of the Forth Bridge.* London: Thomas Telford.

Petroski, Henry. 1995. *Engineers of Dreams: Great Bridge Builders and the Spanning of America.* New York: Alfred A. Knopf.

Proctor, Carlton S. 1934. "Foundation Design for the Trans-Bay Bridge." *Civil Engineering,* December, pp. 617–621.

Purcell, C. H. 1934. "San Francisco—Oakland Bay Bridge." *Civil Engineering,* April, pp. 183–187.

Purcell, C. H., Chas. E. Andrew, and Glenn B. Woodruff. 1934. "San Francisco—Oakland Bay Bridge." *Engineering News-Record,* March 22, pp. 371–377.

Rethi, Lili, and Edward M. Young. 1965. *The Great Bridge: The Verrazano-Narrows Bridge.* New York: Ariel/Farrar, Strauss & Giroux.

Roddis, W. M. Kim. 1993. "Structual Failures and Engineering Ethics." *Journal of Structural Engineering* 119: 1539–1555.

Roebling, John A. 1855. *Final Report to the Presidents and Directors of the Niagara Falls Suspension and Niagara Falls International Bridge Companies.* Rochester, N. Y.: Lee, Mann.

——— 1870. *Report to the President and Directors of the New York Bridge Company, on the Proposed East River Bridge.* Brooklyn: Daily Eagle Print.

Rosenberg, Nathan, and Walter G. Vincenti. 1978. *The Britannia Bridge: The Generation and Diffusion of Knowledge.* Cambridge, Mass.: MIT Press.

Scott, Quintana, and Howard S. Miller. 1979. *The Eads Bridge.* Columbia: University of Missouri Press.

Stackpole, Peter. 1984. *The Bridge Builders: Photographs and Documents of the Raising of the San Francisco Bay Bridge, 1934–1936.* Corte Madera, Calif.: Pomegranate.

Steinman, David B., and Sara Ruth Watson. 1941. *Bridges and Their Builders.* New York: Dover Publications, 1957.

Strauss, Joseph B. 1938. *The Golden Gate Bridge: Report of the Chief Engineer to the Board of Directors of the Golden Gate Bridge and Highway District, California.* San Francisco: Golden Gate Bridge and Highway District.

Tonias, Demetrios E. 1995. *Bridge Engineering: Design, Rehabilitation, and Maintenance of Modern Highway Bridges.* New York: McGraw-Hill.

Trachtenberg, Alan. 1979. *Brooklyn Bridge: Fact and Symbol.* Chicago: University of Chicago Press.

United States Steel. 1936. *The San Francisco—Oakland Bay Bridge.* Pittsburgh: United States Steel.

van der Zee, John. 1986. *The Gate: The True Story of the Design and Construction of the Golden Gate Bridge.* New York: Simon and Schuster.

Waddell, J. A. L. 1916. *Bridge Engineering.* New York: Wiley.

Watson, Philip P. 1987. *The Ambassador Bridge: A Monument to Progress.* Detroit: Wayne State University Press.

Westhofen, W. 1890. "The Forth Bridge." *Engineering,* February 28, pp. 213–283.

Woodward, Calvin A. 1881. *A History of the St. Louis Bridge.* St. Louis: G. I. Jones & Co.

10. Buildings and Systems

Adler, Jerry. 1993. *High Rise: How 1,000 Men and Women Worked Around the Clock for Five Years and Lost $200 Million Building a Skyscraper.* New York: HarperCollins.

American Society of Civil Engineers. *Guide to History and Heritage Progams.* New York: American Society of Civil Engineers.

Anderson, Norman. 1992. *Ferris Wheels: An Illustrated History.* Bowling Green, Ohio: Bowling Green State University Popular Press.

Beedle, Lynn S., editor-in-chief, Council on Tall Buildings and Urban Habitat. 1988. *Second Century of the Skyscraper.* New York: Van Nostrand Reinhold.

Billington, David P. 1983. *The Tower and the Bridge: The New Art of Structural Engineering*. Princeton, N. J.: Princeton University Press.

Building Arts Forum/New York. 1991. *Bridging the Gap: Rethinking the Relationship of Architect and Engineer*. Proceedings of a Symposium. New York: Van Nostrand Reinhold.

Byrne, Robert. 1984. *Skyscraper* [a novel]. New York: Atheneum.

Campbell, Robert. 1988. "Learning from the Hancock." *Architecture*, March, pp. 68–75.

Condit, Carl W. 1960. *American Building Art: The Nineteenth Century*. New York: Oxford University Press.

———— 1961. *American Building Art: The Twenthieth Century*. New York: Oxford University Press.

Cross, Wilbur. 1990. *The Code: An Authorized History of the ASME Boiler and Pressure Vessel Code*. New York: American Society of Mecanical Engineers.

Dibner, Bern. 1950. *Moving the Obelisks: A Chapter in Engineering History in Which the Vatican Obelisk in Rome in 1586 Was Moved by Muscle Power, and a Study of More Recent Similar Moves*. New York: Burndy Library.

Elliott, Cecil D. 1992. *Technics and Architecture: The Development of Materials and Systems for Buildings*. Cambridge, Mass.: MIT Press, 1994.

Escobedo, Duwayne. 1993. "Engineer Shines after New York City Blast." *Engineering Times*, May, pp. 1, 8.

Gavois, Jean. 1983. *Going Up: An Informal History of the Elevator from the Pyramids to the Present*. Farmington, Conn.: Otis Elevator Company.

Giedion, Sigfried. 1947. *Space, Time, and Architecture: The Growth of a New Tradition*. Cambridge, Mass.: Harvard University Press.

Holgate, Alan. 1986. *The Art in Structural Design: An Introduction and Sourcebook*. Oxford: Oxford University Press.

Ishii, Toshiaki. 1994. "Elevators for Skyscrapers." *IEEE Spectrum*, September, pp. 42–46.

Kapp, Martin S. 1964. "Tall Towers Will Sit on Deep Foundations." *Engineering News-Record*, July 9, pp. 36–38.

Kidder, Tracy. 1985. *House*. Boston: Houghton Mifflin.

Kirby, Richard Sheldon, et al. 1956. *Engineering in History*. New York: Dover Publications, 1990.

Le Corbusier. 1931. *Towards a New Architecture*. New York: Dover Publications, 1986.

Levy, Matthys, and Mario Salvadori. 1992. *Why Buildings Fall Down: How Structures Fail*. New York: Norton.

Mark, Robert, ed. 1993. *Architectural Technology up to the Scientific Revolution: The Art and Structure of Large-Scale Buildings*. Cambridge, Mass.: MIT Press.

Marshall, R. D., et al. 1982. *Investigation of the Kansas City Hyatt Regency Walkways Collapse*. Washington, D. C.: U.S. Department of Commerce, National Bureau of Standards.

Morgenstern, Joe. 1995. "The Fifty-Nine-Story Crisis." *The New Yorker*, May 29, pp. 45–53.

Petroski, Henry. 1985. *To Engineer Is Human: The Role of Failure in Successful Design*. New York: Vintage Books, 1992.

Puri, Satinder P. S. 1994. "Trapped in an Elevator During the World Trade Center Bombing: A Personal Account." *Journal of Performance of Constructed Facilities* 8: 217–228.

Robison, Rita. 1989. "Bullwinkle." *Civil Engineering*, July, pp. 34–37

——— 1994. "Malaysia's Twins: High-Rise, High Strength." *Civil Engineering*, July, pp. 63–65.

Rybczynski, Witold. 1989. *The Most Beautiful House in the World*. New York: Viking.

Salvadori, Mario. 1980. *Why Buildings Stand Up: The Strength of Architecture*. New York: McGraw-Hill, 1982.

Templer, John. 1992. *The Staircase: History and Theories*. Cambridge, Mass.: MIT Press.

——— 1992. *The Staircase: Studies of Hazards, Falls, and Safer Designs*. Cambridge, Mass.: MIT Press.

Thornton, Charles H., et al. 1993. *Exposed Structure in Building Design*. New York: McGraw-Hill.

Viollet-le-Duc, Eugène Emmanuel. 1874. *The Story of a House*. Translated by George M. Towle. Boston: J. R. Osgood and Co.

Vitruvius. 1st cent. B.C. *The Ten Books on Architecture*. Translated by Morris Hickey Morgan, 1914. New York: Dover Publications, 1960.

ILLUSTRATION CREDITS

2.1	From Brunel, *The Life of Isambard Kingdom Brunel*
2.3	Courtesy of the Smithsonian Institution
3.1	From Galileo, *Dialogues Concerning Two New Sciences*, 1638
3.2	Courtesy of Staedtler Mars GmbH
3.7	After Koh-I-Noor
3.8	After Cronquist, "Broken-off Pencil Points"
3.9	After Cronquist, "Broken-off Pencil Points"
3.10	After Cowin, "A Note on Broken Pencil Points"
4.6	From Lewis Weiner, "The Slide Fastener," © 1983 by Scientific American, Inc.; all rights reserved
4.7	From Lewis Weiner, "The Slide Fastener," © 1983 by Scientific American, Inc.; all rights reserved
4.8	From *Design News*, February 21, 1994
5.1	Courtesy of *Modern Metals*
5.3	After *Modern Metals*, December 1992
6.1	From Costigan, *Electronic Delivery of Documents and Graphics*
6.2	After Honda et al., "Compacting Technologies for Small Size Personal Facsimile," *IEEE Transactions on Consumer Electronics*, © 1992 IEEE
6.3	After G. Pettersson, "ISDN: From Custom to Commodity Service," *IEEE Spectrum*, June 1995, © 1995 IEEE
7.2	After *Aviation Week & Space Technology*, April 11, 1994
7.3	Courtesy of Boeing Company
7.4	Courtesy of Boeing Company
7.5	Courtesy of Boeing Company
7.6	Courtesy of Boeing Company
7.7	From *Aviation Week & Space Technology*, May 1, 1995

7.8	From *Aviation Week & Space Technology,* January 3, 1994
8.2	From *Illustrated London News,* 1870
8.3	From *New Civil Engineer,* March 14, 1991
8.4	From Viessman and Hammer, *Water Supply and Pollution Control*
8.5	After Steel, *Water Supply and Sewerage,* © 1960 McGraw-Hill
9.3	From *Scientific American,* February 4, 1888
9.4	From Westhofen, "The Forth Bridge," 1890
9.5	From Fowler, *The San Francisco—Oakland Cantilever Bridge*
9.6	From *Engineering,* May 3, 1889
9.7	From *Engineering News-Record,* March 22, 1934
9.8	From *Engineering News-Record,* March 22, 1934
9.9	From United States Steel
9.10	From United States Steel
10.1	From Dibner, *Moving the Obelisks*
10.5	Courtesy of Woolworth Corporation
10.6	Courtesy of Structural Engineers Association of Illinois
10.7	After Ishii, "Elevators for Skyscrapers," *IEEE Spectrum,* September 1994, © 1994 IEEE
10.8	From Robison, "Malaysia's Twins," *Civil Engineering,* July 1994, by permission of ASCE
10.9	From Ishii, "Elevators for Skyscrapers," *IEEE Spectrum,* September 1994, © 1994 IEEE
10.10	After Chang and Yang, "Control of Buildings Using Active Tuned Mass Dampers," *Journal of Engineering Mechanics,* March 1995, by permission of ASCE
10.11	From Thornton et al., *Exposed Structure in Building Design*
10.12	Courtesy of Structural Engineers Association of Illinois
10.13	Courtesy of Port Authority of New York and New Jersey
10.14	From *Engineering News-Record,* March 8, 1993

Page numbers in italics refer to figures and their captions.

ergonomics, *see* human factors

errors, in analysis, 59–60, 63. *See also* Galileo, error of

ethics, 163

Everett, Washington, 129

evolution, of artifacts, 15, 205–6

experiments: demonstrating failure modes, 93, 95; unplanned, 55

Exposition Universelle (1889), 197

extrusion process, for plastic bags, 84, 85, 86

F-117A (airplane), 132

facsimile, *see* fax

factors of safety, 65

failure, 89, 90, 92–103 *passim*, 104; analysis, 54, 121, *121*, 147; criteria, 4, 89–91, 96, 98; environmental, 90, 98–99, 101, 103; modes, 95

falsework, 160, *161*

Faraday, Michael, 147

Fastener Manufacturing and Machine Company, 70

fastening devices, 66–88; patents for, 66–88 *passim*. *See also specific devices*

Fasullo, Eugene, 212–3

fatigue, metal, 121–3, *121*

fax machines, 5, 105–118; Asian markets for, 113–4; compatibility of, 110, 114; cost, 115; digital, 114; early, 106–8, *117*; Group 1 (standard), 113; Group 2 (standard), 113; Group 3 (standard), 114; Group 4 (standard), 118; Japanese, 113–4; personal size, 108, 116–7, *117*; scanning of images by, 106–8

fax transmission, 106–10; of newspapers, 108, 110; standards for, 112–3, 114; 118; telephone lines and, 109–10, 116; time of, 113, 114, 118

Federal Aviation Administration (FAA), 135, 137

Federal Communications Commission (FCC), 110

Federal Express, 115

ferries, 168. *See also* San Francisco Bay, ferries

Ferris Wheel, 204

fiber-optic cables, 138

filtration, of water supply, 156, *156*

Firth of Forth, 168, 169, 170

Firth of Forth Bridge, 169–71, *170*, *173*, 180, 183

Firth of Tay, 168. *See also* Tay Bridge

Flatiron Building, 194

Flexigrip (plastic zipper), 81–2

fluid mechanics, 159

forces: in cantilever beams, 46–7, *47*; components of, 48, *49*; on pencil points, 48–62 *passim*; shear, 58, 60–1, 62. *See also* stress; wind, force of

form vs. function, 27, 29

Forster, Henri, 72

Fort Lee, New Jersey, 164

Fort Pierce, Florida, 101

foundations, 179, 198

Fourth Street Bridge (San Francisco), 165

Fowler, Charles Evan, 167, 168, 171–5, *172*, 183

Fowler, John, 169, 170

fracture, *see* pencil points, broken-off; fatigue, metal

Fraze, Ermal, 93, 94, *101*

friction, 48

Froelich, Linda A.,

Froelich, Richard D., 39, 40, *41*, 42

Frontinus, 142–3, 145

Fuji (Japanese company), 130

Galileo, 2, 45, 46–7, 52, 59; and cantilever beam, *44*, 45–65 *passim*; error of, 10, 47, *47*, 58, 63

Galloway, John D., 177

"Gang of Eight," 125, 126

Gem (paper clip), 3, 8, 16–39 *passim*, *17*, 24; aesthetics of, 24; faults of, 19, 24, 25, 29–30, 32–3, 34, 39, *41*; improvements on, 18–42 *passim*; origin of, 16–8; praise of, 27, 29; proportions of, 19, 24–5, *24*, 27, 29, 39. *See also* paper clips; *specific clips*

Gem Limited (firm), 18

General Electric Company, 135

George Washington Bridge, 164, 175, 179, 182

Gilbert, Cass, 198

girders, 191

glue, used in pencil making, 52–4